CAMBRIDGE LIBRARY COLLECTION

Books of enduring scholarly value

North American History

This series includes accounts of historical events and movements by eye-witnesses and contemporaries, as well as landmark studies that assembled significant source materials or developed new historiographical methods. The works range from the writings of early U.S. Presidents to journals of poor European settlers, from travellers' descriptions of bustling cities and vast landscapes to critiques of racial inequality and descriptions of Native American culture under threat of annihilation. The commercial, political and social aspirations and rivalries of the 'new world' are reflected in these fascinating eighteenth- and nineteenth-century publications.

Notes on the Mineralogy, Government and Condition of the British West India Islands and North-American Maritime Colonies

The most renowned naval officer of the mid-nineteenth century, Thomas Cochrane, Tenth Earl of Dundonald (1775–1860), served in wars against Spain and France, retiring as an admiral in the Royal Navy. He was also an M.P., vociferously calling for naval reform in Parliament. Due to a financialscandal, he left the Royal Navy for a period and became a celebrated mercenary, commanding naval forces in the wars of independence of Chile, Peru, Brazil and Greece. First published in 1851, this work contains notes on a voyage of 1849 around the West Indies and North America. Describing the peoples and geography encountered, it offers progressive remarks on the end of slavery, criticisms of plantation owners, and suggestions for commercial improvements. The book remains of enduring interest to scholars of naval, colonial and Caribbean history.

T0188094

Cambridge University Press has long been a pioneer in the reissuing of out-of-print titles from its own backlist, producing digital reprints of books that are still sought after by scholars and students but could not be reprinted economically using traditional technology. The Cambridge Library Collection extends this activity to a wider range of books which are still of importance to researchers and professionals, either for the source material they contain, or as landmarks in the history of their academic discipline.

Drawing from the world-renowned collections in the Cambridge University Library and other partner libraries, and guided by the advice of experts in each subject area, Cambridge University Press is using state-of-the-art scanning machines in its own Printing House to capture the content of each book selected for inclusion. The files are processed to give a consistently clear, crisp image, and the books finished to the high quality standard for which the Press is recognised around the world. The latest print-on-demand technology ensures that the books will remain available indefinitely, and that orders for single or multiple copies can quickly be supplied.

The Cambridge Library Collection brings back to life books of enduring scholarly value (including out-of-copyright works originally issued by other publishers) across a wide range of disciplines in the humanities and social sciences and in science and technology.

Notes on the Mineralogy, Government and Condition of the British West India Islands and North-American Maritime Colonies

Thomas Cochrane

CAMBRIDGE UNIVERSITY PRESS

Cambridge, New York, Melbourne, Madrid, Cape Town,
Singapore, São Paolo, Delhi, Mexico City

Published in the United States of America by Cambridge University Press, New York

www.cambridge.org
Information on this title: www.cambridge.org/9781108054065

© in this compilation Cambridge University Press 2012

This edition first published 1851
This digitally printed version 2012

ISBN 978-1-108-05406-5 Paperback

NOTES

ON THE

MINERALOGY, GOVERNMENT AND CONDITION

OF

THE BRITISH

WEST INDIA ISLANDS

AND

NORTH-AMERICAN MARITIME COLONIES:

WITH

A Statistical Chart of Newfoundland,

CONTRASTING THE

CIRCUMSTANCES OF THE FRENCH AND BRITISH FISHERIES.

BY

ADMIRAL THE EARL OF DUNDONALD, G.C.B.

LATE NAVAL COMMANDER-IN-CHIEF ON THE ABOVE STATION.

LONDON:

JAMES RIDGWAY, PICCADILLY.

1851.

PRELIMINARY REMARK.

REPORTS of Committees of Privy Council on trade since the peace of 1783, and of Committees of the House of Commons since the termination of the last war—together with less remote proceedings of the Legislature—prove that colonial distress cannot be remedied by Authorities less powerful than those who sanctioned or enacted the measures that produced it. It is not therefore to Colonial Governors abroad, or to Colonial Secretaries at home, that we are to look for the removal of evils.

PREFACE.

HAVING availed myself of my recent opportunity
of becoming acquainted by personal observation
with various particulars connected with the Natural
History, Colonial Government, and social position
of the British possessions throughout the extent of
the Naval command in North America and the
West Indies; and having taken Notes of all such
matters as appeared to me most interesting, I
commit them to the press in the hope that they
may attract attention in our own country; for the
impression on my mind is, that were the people of
England acquainted with the condition of the Colonies in question, and conscious how valuable and
important they might become, not only in a commercial point of view, but as a source of power to
the Parent State, instead of remaining, as they now
are, causes of expense and weakness; they would,
with the irresistible voice of unanimity, call upon
the Legislature to award justice to our hapless and
helpless fellow-subjects in those interesting possessions. The notion entertained by some that Colonies
are valueless and even injurious to the Parent State,
is true or false according as they are ill or well
governed. Deliver them from misrule, and concede

to them their just share of natural rights, and their influence for good would be felt to pervade the whole of the British dominions.

It was well to grant twenty millions for the emancipation of slaves throughout the Colonies. It had the happy effect of imparting joy to the hearts of a multitude of oppressed and desponding beings, and would have been productive of unalloyed good, had not the Planters generally been in debt to the whole extent of their proportions of that vast amount, in consequence of their having kept up the same extravagant expenditure as before their incomes had become diminished by various well known causes. Hence, when deprived of slave-cultivation, few had the means of defraying the cost of free labour. The evil of inability to pay the emancipated negro population adequate and constant wages, has operated to the present day in continuing and increasing their reluctance to work, and in perpetuating the belief of their love of idleness so usually laid to their charge. But how can it be expected that the negro will labour for an inadequate or uncertain pittance, or that the Planters can now afford to pay for cultivation under the additional infliction of that anomalous free-trade policy which so inconsistently encourages the production of slave-grown sugar by other nations, at the same time that half a million a-year is expended in the maintenance of a squadron for the suppression of the odious slave-trade! Were that sum applied to the relieving of the Colonists from the burthen of supporting the disproportionate governmental establishments existing throughout the Colonies, as far as it would go—it would do something towards enabling the Planter to pay for free labour—something towards enabling him to compete with his slave-holding rivals, who probably pay no such taxes—and do more towards the suppression of the slave-trade than it does in the way in which it is at present applied. But while every possible burthen is imposed, in addition to a land-tax of some ten shillings for every cultivated acre, and eight shillings on every hogshead of sugar embarked, there can be no hope of the return of prosperity to our sugar-growing Colonies.

Strange as it may appear, it is unquestionably true that the practice of taxing the necessaries of life, which has been relinquished at home, is unrelentingly pursued throughout these ruined Colonies. Flour, and every kind of food imported, and even articles of native produce, are rigorously taxed. By this system the poorer classes are still further impoverished, and at the same time are prevented from obtaining small lots of land (whereby to support themselves) by reason of the heavy charges imposed for title-deeds to small holdings, and imposts laid on their produce. Were it not for these exactions, even the poorest labourer might acquire a small property in the soil when abandoned estates are sold at almost a mere nominal value. The unwillingness to admit the labouring classes into the possession of small portions of land is partly the effect of an apprehension that labourers, having land of their own, would not be available for the cultivation of the principal estates; but there seems no more reason why this should be the case in the Colonies than in England; where, as I am informed, cottagers and those who have allotments of land for spade-husbandry (which they exercise at over-hours) are the most willing and effective labourers employed on the adjacent farms.

But—melancholy it is to relate—thousands of these unhappy negroes, having neither occupation of their own, nor employment imparted, are reduced to the lowest depth of destitution; and shocking it is to behold the poor ragged creatures, whose dirt and diet invite cholera and pestilence, lying about in a state of inactivity and mental abandonment; espepecially in the once busy sea-ports, now generally shunned by reason of the inordinate imposts which deter vessels from entering, except those whose expenses are defrayed by the local planter. The noble Port of Kingston, in Jamaica, though better situated as an intermediate station than the free and crowded Port of St. Thomas's, is deserted—save when a solitary steamboat arrives for the purpose of obtaining coals, invited, in that case, by a relaxation of fiscal exactions. The like ruinous policy is practised throughout the whole of the West India Islands.

It would be gratifying could I pass from the

deplorable condition of the proprietary and population of these islands, consequent on the injurious
enactments and exactions there enforced, to a better
state of things in the Northern Provinces, which,
were their resources rightly developed, and the tide
of emigration directed thereto by the promulgation
of a sound and liberal system of colonization, would
form a possession as advantageous as creditable to
the British Crown. The soil, it is true, and its
power of production, can only come in competition
with those of the United States, by reducing the
colonial and commercial imposts, and so leaving in
the hands of intelligent settlers the means of clearing
and cultivating the tracts of which they become
possessed : a remedy within the power of the British
Legislature, and happily on a principle of inter-
ference directly the reverse of that which severed
from us our former transatlantic Colonies.

The vast mineral wealth of our still remaining
American possessions is yet comparatively unde-
veloped. In Nova Scotia scarcely more than one
superficial mile of coal out of an extent of several
hundreds of superficial miles has yet been excavated,
and numerous mountains of iron ore exist, so pure,
as to be available for its purposes with little previous
preparation ; one of which mountains only is in an
incipient state of production. It is an extraordinary
fact, that no mineralogical survey on the part of
the British Government has ever taken place, and
that the only certain knowledge of the mineral

riches of Nova Scotia and New Brunswick is derived from the labour of the learned and indefatigable Dr. Abraham Gesner, who has published several works of high national importance on the mineralogy and resources of these Provinces. The American Government employed a scientific person on their behalf to investigate and report upon these interesting objects, which he did, but whose report I have not seen; and I met a French officer last year at Sydney, travelling for a similar purpose.

When it is considered how important the Coal mines,—situated on the margin of the finest ports in Nova Scotia and Cape Breton, are, in a naval point of view, and that the various and available resources of these possessions are unrivalled throughout the whole extent of the southern shores of North America, it seems astonishing that no care has been taken to secure a respectable body of emigrants, of undoubted loyalty, to occupy these interesting Provinces.

Would it not be highly creditable to young gentlemen of influential families—instead of regarding the Colonies as substitutes for the monastic institutions of former times, and seeking sinecures in them at the expense of the community—to shew the virtuous example of resorting to and locating in these Provinces, where, by assiduously engaging in the pursuits which Nature has there so amply provided, they could not fail to acquire the honourable means of enriching themselves, and becoming useful members of society ?

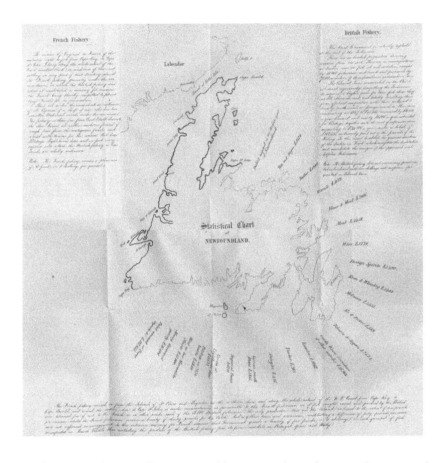

The material originally positioned here is too large for reproduction in this reissue. A PDF can be downloaded from the web address given on page iv of this book, by clicking on Resources Available .

NOTES

ON THE NORTH AMERICAN AND WEST INDIA COLONIES, &c.

WE left Halifax early on the morning of the 14th July, 1849. The wind being contrary and light rendered it necessary to beat out of the harbour. Having effected that object, the Wellesley proceeded in a course as nearly along shore as the numerous known and reported rocks would permit. The shore was at such a distance, that its capability for improvement, and the nature of the rocks of which it is composed, could not be distinctly observed. It seemed almost uninhabited, and there is little doubt of its being equally unproductive, and of the same material as the soil and strata about Halifax.

We passed several schooners employed in the coasting trade, and a considerable number of fishing boats during our progress towards Louisburg, which we reached on the morning of Tuesday the 17th. The entrance of this harbour is between a light-house on the right hand side and a small island on the left (once well fortified.) The clear course is mid-channel, leaving a rock on the starboard side which projects about one-fifth into the channel. The harbour of

Louisburg is quite safe for vessels of any size, and is capacious enough, both to the north-east and south-west of the channel, to contain a large fleet. This is an admirable port whereat to carry on an extensive fishery. There are at present only thirty-two small boats, each having two men.

Although the destruction of Louisburg after its capture may have been politically necessary, yet it is lamentable to see the wrecks of these costly fortifications, and (with the exception of five huts) weeds and grass growing where once stood the flourishing city of Louisburg, having a fishery more extensive than all that remains to England, under the retributory measures adopted by France in the awarding premiums, to ensure a successful competition with British fisheries.

How different are the views now taken of this important nursery for British seamen, and of this commerce so jealously watched and fostered in former times!

On the 19th the Wellesley weighed and proceeded towards Sydney, and at daylight on the following morning rounded Low Point, and at three P.M. anchored off the Coal Mines, where there are a number of vessels loading with coal. The coal here is said to be far preferable to that of Picton, both for steam vessels and for common purposes. The Americans use it for the manufacture of gas, but they prefer the Picton coals, (it being free from sulphur,) for the smelting of iron.

There are no fishing boats out of the Port of Sydney, I believe not even for the supply of the town, as we could not procure any fish.

It is in the coal measures here that the wonderful curiosities of trees, roots, branches, and ferns, are found beautifully preserved, exhibiting even the most minute ramifications in the foliage and seeds of plants which existed previous to the deposition of 1300 perpendicular feet of distinct stratified deposits. Where the oblique minerals crop out at this depth the fossil trees seen in the cliffs, on the sea shore, are vertical; many of them have their roots entire, and from the top to the bottom pass through several strata of different materials, and in some places reach from a bed of coal of one foot to a superincumbent coal of four feet in thickness.

The very intelligent agent of the colliery has a beautiful collection of the most curious objects, some of them of great rarity.

On the 23rd of July we weighed and stood along shore towards the north, passing the Channels into the Bras d'Or, a magnificent lake which occupies the whole centre of the Island of Cape Breton. We passed the low flat islands and rocks off its entrance, at some distance on our left, not knowing exactly how near we could safely approach, as the survey now in progress has not yet been published. At evening we shortened sail, and on the following morning found ourselves off Neganish Bay, the whole district presenting a

barren and desolate appearance. Thence we proceeded towards Cape North and got close in shore,
the mountains being precipitous and the contiguous
water deep. Here we landed in the gig, and
brought off some specimens of the numerous strata
which stand nearly perpendicular; after which we
bore up for St. Paul's Island, and passed on the
eastern side. Here there is a cove, in which the
lighthouse keeper resides, with some good-looking
pasture around it. The rocks stand nearly in a
perpendicular direction, and appear to be chiefly
of igneous origin, but near the little detached island
on the northern end, strata, seemingly micaceous
slate, appear. I have seen some very rich specimens of lead ore, said to have been picked out of
these heterogeneous materials. A heavy surf rolled
along the beach, therefore we did not attempt to
land, but bore up towards Cape Rouge.

NEWFOUNDLAND.

On approaching this Cape, off which there is an
island forming an anchorage, we saw therein a
schooner, which proved to be that of the Bishop of
Newfoundland, on his visit to the various detached
settlements, between which there are no practicable
roads. Captain Goldsmith went on board to pay his
respects to the Bishop, but found he was on shore.
Here we saw the French flag flying on several fishing vessels.

A river runs down from the mountains in a ravine
close by this anchorage, on the banks of which
enormous blocks of stratified stone of a diagonal
form, seemingly of a micaceous slate, dipped towards
the S.W. at an angle of about 80°. The river is
called the Great Codroy River, situated south of
Cape Aguille, on the northern side of which Cape,
the strata inclines about the same angle towards the
N.W. The diagonal sheets of stratified stone split
in a S.W. and N.E. direction. There are places, a
few miles to the N.E. of Cape Aguille, where
volcanic matter crosses the stratified rock; it seems
probable that metallic ores may exist here, but the
strata are not carboniferous, although it is reported
that coal is found at some small distance in the
interior.

The high land of Cape Aguille ceases about First
River, and is succeeded by low grounds, the upper
part of which is alluvial, and the lower portion
seems to be of old red freestone almost on edge,
being nearly at the same angle, and split in the same
direction as the micaceous rocks on the north side
of Cape Aguille. The colour of this stone is more
vermilion than red.

We observed a number of farm houses on this
alluvial soil; the fields were greener and the flocks
more numerous than at any place we had before
seen, although on this portion of Newfoundland
British subjects are prohibited from settling; and by
treaty, the subjects of France were not to have any

establishment, but were merely to resort to its bays and harbours for the purpose of curing their fish. Being doubtful, from the imperfect charts of this coast, whether the Wellesley could enter St. George's Harbour, we anchored to await daylight, and on the morning of the 27th July, we shifted nearer to the port, and the Master, who was sent to examine the anchorage, having found it capacious and safe, the Wellesley proceeded; and, in truth, a finer port is scarcely to be found. A low spit of land protects it from the sea. This spit is precisely of the shape and nature of that called the " Palisades" at Port Royal, Jamaica, but the harbour is preferable in all respects. Here again we found the Bishop's schooner at anchor, having arrived in the night, in furtherance of his laudable endeavours to visit all his cures once a year, in which his time is almost wholly occupied, and a large portion of his revenue expended. The Captain went on board the Bishop's vessel, and having inquired at what hour it would be convenient I should pay my respects to him, his Lordship appointed eleven o'clock to receive me. He shewed me a chart of the coast of Labrador, whereon was marked the track of his vessel last year, as far north as Sandwich Bay. He presented Captain Goldsmith with a printed copy of his Journal, and I ascribe to his diffidence his not having offered one to me. I read the document, however, with great interest, but I should have felt more satisfied had he given an account of the characteristics of Labrador, and its general appearance at the ports he entered. One fact noticed by his Lordship at the one hundredth page of his Journal struck me forcibly, namely, that the inhabitants of English Harbour had been prevented from completing their little church and school-room (commenced last year) by a tax on herrings in bulk, by which their slender means were so straitened, that the difference in their circumstances since his last visit was very visible.

To return to St. George's Harbour. It will be seen, by reference to the chart, that it is situated in the angle of a deep bay, between Aguille and Cape St. George, the town being on the promontory, and having deep water close to it. No village can be better placed for the herring fishery, as these gregarious fish, at the season of their arrival on the coast, enter this harbour, as it were into the cod of a net, whence they are lifted into the boats by scoops and buckets. With such slender means possessed by the inhabitants, the average catch amounts to 22,000 barrels, but hundreds of thousands might be taken, were encouragement afforded. Salmon is also caught in the neighbouring rivers, which are alive with undisturbed neglected trout. The barrels in which the herrings are packed, are said to cost 2s. 6d. each, and some new regulation requires additional hoops, which, to those concerned, appears a grievance. It is said the herrings must realize 10s. per barrel, in order to repay cost and labour; but the last

advices from Halifax state that 8*s*. only are offered
by the merchants.

The French, I understand, attend more to the
cod fishery. They are not at liberty, if they adhere
to the Treaty, to draw nets on the shore. There is
an American merchant here who deals in truck
with the English settlers, and obtains from them
about a third part of the herrings caught, which he
sends to the United States, in such of the numerous
American schooners employed in the fishery as
enter this bay.

The unauthorised British settlers here, are said to
be very jealous of intruders, as they consider they
have an exclusive right to the land and fisheries in
their actual possession, and from which all are by
treaty excluded. They seemed suspicious that the
Wellesley might have some motive in entering the
bay, contrary to their interests. No person whatsoever came on board, nor did any one come off to
the ship, even to offer himself as a pilot. Some persons were lately desirous to set up a saw-mill, which
would have been important, as they obtain all their
staves for herring casks, &c., from abroad, but the
sanction of the inhabitants could not be obtained.

There is no Magistrate, or Civil or Military
authority, no Medical man, and perhaps fortunately,
no Attorney. Indeed there is no law, though justice is done amongst themselves after their own
manner. There is a neat little church, at which
the Bishop is now officiating, and the people who

are resorting to it seem well dressed and orderly.
I am convinced that in this state of society it is
better to leave the fruits of the earnings of individuals to fructify in the prosecution of their humble
avocations, than after having taxed cottages, sheds,
gardens, and all they possess, to lay a duty (as in
the case mentioned by the Bishop before alluded to)
on herrings in bulk ! Here there are no port dues,
or other similar exactions, nor payments for clearances, which grievously burden and delay such small
commercial transactions.

At daybreak on Monday, the 30th July, we
weighed anchor and stood along the peninsular
promontory called Cape St. George. About midway
a remarkable change takes place to the northward
of the Table Mountain, where the vertical strata
become in appearance horizontal along the whole
shore of the projecting isthmus. The colour of the
strata is chiefly grey, in parallel layers of varying
hardness, as appears from its projections and inden-
tations. I could not, without delaying the ship
longer than I wished, procure samples of the strata,
but there was no appearance of carboniferous
minerals. The same layers were visible in detached
places up to the tops of the hills which are of considerable altitude, though that is not denoted in the
chart.

When we rounded Cape St. George on the following morning, the strata which before appeared
parallel was observed to dip at a considerable angle

towards the north-east, and seemed, where sufficiently exposed to view, to be split into large diagonal flakes, like those noticed before our arrival at St. George's. There is an island close off the shore, about five miles to the eastward of the Cape, called Red Island; which is of quite a different formation, seemingly red horizontal layers of sandstone, of a soft nature, as is obvious from the encroachments of the sea. The peninsula opposite to this island is of considerable elevation, as far as Round Head, whence it gradually lowers to a point about ten miles farther to the eastward. Here the level ground at first seems to be alluvial, but on closer observation, indurated igneous rocks are seen to protrude in flakes, dipping into the sea.

The bay formed by this promontory is of great magnitude. There are several islands at its mouth and in the interior, but there being no chart and no motive for entering it, we stood on towards the mountains on the main shore, some of which are very high. In many parts the contortions of the strata, and the confusion of all kinds of materials, are extraordinary. We intended to anchor in the Bay of Islands, but on arriving off that inlet, a very high precipitous mountain on the right hand of the entrance completely shut off the wind, and as there was a risk of our drifting about in the bay all night, in the outer part of which there is no anchorage, we stood towards an enormous mountain which appeared through the hazy atmosphere.

On crossing this bay we had a more extensive view into the interior, where we observed the bare heads of several white coloured mountains, precipitous parts of which appeared like gypsum.

During all this day's run the sides of mountains on the shore are clad with moss alone; trees, of very stunted growth, only appearing in the sheltered valleys. No visible portion of the shore seems capable of producing food for man; a bright verdure exists indeed in patches like luxuriant meadows, but these are covered with moss, and I believe from the climate, incapable of cultivation.

At six o'clock in the evening we were off Trout River. The boat was sent to a contiguous bluff between that and Port-a-Port, and brought off some specimens of the rock which was of a nature I never saw before. It will doubtless be rightly specified by the Institution of Practical Geology, to which I have presented all my boxes of minerals.

The ship was hove-to during the night, with a view to run along the coast in the morning, and thereby form an estimate of its nature and capabilities.

August 1st.—At daylight we found ourselves eight miles to the north-east of Trout River, and immediately off the lofty precipitous mountain which forms the western entrance of Bonne Bay. The general character of the rock seems the same as of that at Trout River, having, as in other parts, irregular veins traversing in former fissures which render it probable that metalliferous deposits exist there. I did not feel justified, however, in detaining the ship for a cursory inspection. All the coast we had passed invites the research of a Geologist. What the importance of the minerals in the interior of this island may be it is impossible to conjecture, but at the distance of about twenty miles inland there are three enormous mountains over-topping all the intermediate hills of great altitude.

A few miles to the north-east of Bonne Bay, Green Point projects from the shore, exhibiting in its perpendicular face stratified rocks of various kinds, dipping at an angle of 70° or 80° into the sea. The projecting point is formed of a conglomerate of disintegrated and crushed materials, evincing the vast force that had operated to produce its elevation. There is a great tract of low country (or rather flat surface), reaching far into the interior, which possibly may be carboniferous strata, upheaved by the vast masses of igneous rock which constitute the surrounding mountains.

The weather had now become so hazy, that the character of the strata in these plains could not be ascertained without landing, nor indeed would the nature of the stunted brushwood permit any progress to be made in the limited time we could spare. We hove-to, to await the clearing of the weather, and saw some fishing boats, amongst which there was the boat of a poor old man and his son from Trout River, who told us they were the only inhabitants in that quarter, his wife and several children being dead. He said they had buried them, and that his son would probably perform that office to him. He was born in Dorsetshire. They had both the appearance of living wholly on fish, by their vermilion thin-skinned complexion, bleary eyes, and spare figure. No vessel approaches their sequestered abode, as there is no shelter at the mouth of the Trout River.

The fog having cleared away about noon, Cow Head was discovered, a promontory extending some miles into the sea, forming two bays and a small harbour for fishing boats. The wind being on shore, we could not anchor with prudence, though I wished to have done so in order to obtain information about the French fishery, which is said to be carried on to a considerable extent. Here the mineral strata render the existence of coal probable in the surrounding country. Between Seaford River and Sandy Bay some immense acute-angled blocks of a reddish-coloured stone appear in various places, and a considerable stream of water falls over a rock, probably of sandstone, from the appellation, Sandy Bay, given to the coast where the river empties itself. The bluffs projecting into the sea which form this bay, are alluvial deposits, and seem fast washing down, which may account for the usually rocky bottom being soft off this place, to the depth of twenty-five fathoms.

Portland Hill stands alone upon this plain contiguous to the beach. It is composed of a bed of deposits, apparently as if formed by showers of cinders and other matters of a volcanic nature, though I do not believe it to be of volcanic materials. The general colour is grey, and the strata are evidently of different hardness as they project irregularly. I felt anxious to ascertain the nature of its strange appearance; but the matted brushwood, which will not burn, afforded no prospect of accomplishing that object in a reasonable time.

To the eastward of Portland and at the southern extremity of the plain, about twenty miles distant, there is a range of broad and lofty mountains, at least ten or twelve thousand feet in height, which I believe can be approached by a river that traverses this low plain.

Assuredly it would well reward a geological and mineralogical research both as regards information and pecuniary advantage. In passing the mouth of this River of Ponds, there was a wreck upon the beach, probably a timber ship, which we hauled in to assist, if necessary ; but as no signal was made, nor did any boat come off, though there were several on shore, we bore up.

The strata on the shore are apparently indurated deposits of clay, or rather mud, the hardening of which may have been connected with the occurrence of the granite, which appears in some places at the water's edge. The land, though level above this seemingly baked clay, was almost destitute of soil.

From Mal Bay, into which the River of Ponds
flows, we ran along the coast without any thing
presenting itself worthy of notice ; and at dusk on
the evening of the 1st August, when abreast of
Point Rich, we stood across the straits about N.E.,
towards Wood Island, on the Labrador coast;
but the wind failing, we found ourselves in mid-
channel at daybreak. The fog afterwards partially
cleared up, and the land on both sides became visible·

LABRADOR.

Towards sunset we got in with the Labrador
shore, on one of the most beautiful evenings which
could any where be seen. The wind having changed,
we did not get into Forteau Bay that night, but on
the morning of the 3rd August we anchored there—
the brilliancy of the sky, and the weather being
such as would have done credit to the Mediterranean.
In a contiguous harbour, called Port Modeste, there
were numerous British vessels of considerable size.
In Schooner Cove a brig and schooner were at
anchor ; and in Forteau there was a barque, a brig,
and a schooner. The small boats at the mouth of
the bay have seemingly an excellent fishing ground.
The people are all hard at work. No sooner is one
line hauled up than the second requires attendance.
The fish flakes are numerous on the beach, and so
are the men turning the fish, although so many are
out in boats. I counted forty-eight men hauling up
a fishing boat by main force, there being no crab,

windlass, or tackle employed. There are heaps of
fish cured and stacked up, like diminutive hayricks,
the fish being nicely arranged to throw off the rain,
like thatch. I suppose this is done to save the cost
of a greater number of storehouses, which here are
all built of sawn boards; logs not being used as in
Nova Scotia and Cape Breton, where wood is plentiful. The valleys here, as well as the hills, as far
as we could see inland, are destitute of trees of any
kind higher than gooseberry bushes; these growing
only in the hollows, less exposed, have the appearance of hedges, whilst the intermediate parts,
covered with yellow withered grass, seem to the
view like English corn fields in autumn.

The absence of vegetable productions of sponta-
neous growth is by no means a proof of sterility in
this region as to horticultural produce. In a garden
here we saw potatoes of excellent quality growing
from others of last year's crop, produced on the
spot. The owner of the garden shewed us lettuce
and radishes as forward as those we left three
weeks ago in the Admiralty garden at Halifax.
I find a wild pea growing along the beach, the
stalks of which our sheep eat with avidity; there
being no seed on them at this time of the year, I
have taken up as many roots as I can carry, to try
if they can be cultivated in the manner of vetches;
if so, it would be a great boon to Nova Scotia,
where fodder is scarce. I have likewise procured
plants of another wild vegetable, which the sheep

eagerly devour. Perhaps a botanist might discover useful plants in this region adapted to English agriculture.

There is no convenient place for watering ships at Forteau, although there is a waterfall on the west side of the bay, and a river at the top in which some of the officers had great success in trout fishing.

August 4th.—At 9 A.M., when the tide began to run to the eastward, we weighed anchor, and made sail along the Labrador shore, towards Red Bay, in the Straits of Belleisle, and passed several fishing stations, apparently snug harbours, containing numerous schooners, and also brigs and larger vessels, which transport the fish caught on the Labrador coast to England. The activity apparent here in the fishing department I ascribe to the British fishery on the coast of Labrador being still free from the exactions now imposed in Newfoundland, for the maintenance of the excessive retinue of a Government, by the taxation of every article of food and clothing imported, together with imposts placed on all existing property, none of which is of a productive nature.

The strata from Forteau to Black Bay are all horizontal, and of the same nature as the specimens procured at the former place ; but from Black Bay to Red Bay volcanic rocks are interposed in various places. Some dark grey-looking strata, in parallel lines puzzled me as to their origin, but I do not think that coal, or any valuable minerals exist in this neighbourhood. But the British possessions in Labrador extend over a tract of country as great as the northern regions of Russia from St. Petersburgh towards the Pole, wherein the Ural Mountains compensate that Government for the sterility of the soil. I have often felt surprised at the indifference evinced by the Spanish Government as respects the development of the resources of their possessions; but it is with still greater astonishment that I view the supineness of our own Government in leaving this vast tract unexplored, and its probable treasures undiscovered.

NEWFOUNDLAND.

From Red Bay we crossed over to the east end of that part of Newfoundland called the Strait Coast, by reason of there being no harbour or inlet in the whole extent. I had no opportunity of getting a piece of the horizontally stratified rock exhibited every where along shore, but I am satisfied that it is all primitive sandstone, like that obtained at Forteau.

After passing Boat Head volcanic products overlay the primitive formations. The strata in many places are curved in a curious manner.

Continuing our course to the eastward, we soon got abreast of the islands called, on the very imperfect English charts, Sacred Islands, but which, I believe, to be all erroneous translation of a slang

French expression; the whole being dark, volcanic substance, without a tree, bush, or green thing, existing thereon. Cape North is also of the same material, having imperfect basaltic columns almost perpendicular. We had an excellent view, by reason of our contiguity, and the rays of the setting sun full upon the objects. The nature and origin of this Cape were so plain, that I did not think it worth while to detain the ship in order to procure a specimen. We passed Belle Isle at some distance, but it has precisely the same character as Cape North.

Being now outside the Strait we saw five icebergs, on one of which was a large animal, probably a walrus. Two islands were also in sight, called White Islands, because the rock is of a much lighter colour than Cape Bauld; but I do not believe it differs except in colour. We continued our course alongside towards the southward, passing several ports wherein there were numerous French ships and square rigged vessels dismantled, and schooners and multitudes of fishing boats in full activity in the offing. These schooners and fishing boats are manned by the crews of the large French vessels which are laid up in port, and constitute depôts as well as the means of transporting the produce of the fishery to France, an arrangement highly advantageous to the French marine, and which we erroneously abandoned by erecting Newfoundland into a Colonial Government—thus surrendering our deep sea fishery entirely—even without rendering the inshore fishery available to the newly erected Colony, throughout which it languishes from want of stimulus, or an adequate reward, even to induce the impoverished inhabitants of the shore to avail themselves of their small and almost costless boats to catch fish, which, by reason of the bounties given by France and America, are unsaleable with profit in any country in Europe.

It is grievous to observe the difference in the mode of carrying on the British fishery compared to that of the French. The former in rudely constructed skiffs with a couple of destitute looking beings in party-coloured rags; the latter in fine well equipped schooners (which may be called tenders to their larger ships), the seamen uniformly dressed in blue with Joinville hats, looking as men ought, and may be expected to look, whose interests and those of the Parent State are understood to be in unison and attended to as such.

We regretted much that the wind having changed directly on shore prevented our entering Croc, the chief station of the French fishery, because of the difficulty we should have experienced in getting out of that narrow harbour without the aid of a steam vessel. The same reason prevented our inspecting the British fishing stations on the east side of Newfoundland; so the course was directed to clear all dangers in our route to St. John's. We passed close to a rock or isle of grey granite, bearing the ominous appellation of Funk Island, whereon there was not the slightest appearance of vegetation, and then successively rounded the promontories called Capes Freels, Bonavista, Breakheart, and St. Francis, between which there are deep bays, all exhibiting the most forbidding appearance—not a vessel was seen on the whole run, not even a boat.

We arrived off St. John's on the 8th August and entered the harbour, which is spacious in the interior though narrow between the Capes. The town presents an agreeable appearance after the desert parts we have seen. The opposite or western side of the harbour is covered with scaffolds for drying fish, at which there were numerous persons occupied in turning them over, in order to their being dried by the wind.

The Governor's Aide-de-camp came on board with a very polite invitation, which I accepted; and having proceeded to Government House, His Excellency Sir Gaspard Le Marchant was most courteous; and on my expressing a desire to obtain information on the subject of the fisheries, and the general state of the Colony, he not only offered to place all official sources of information at my service, but recommended my taking up my abode at Government House, in order to facilitate the object by the inspection of official documents in the office.

Under existing circumstances, His Excellency has most zealously, energetically, and usefully, directed his attention to free the population from the odious oppression of the truck-system, by encouraging the fishermen to cultivate vegetables for their own use, and has set an example of farming on the Government grounds to some extent, offering premiums to those who successfully follow agricultural pursuits.

This, I truly believe, is the utmost he can effect, either as an individual or in his official capacity ; but it can in no degree avert the effects of the heedless policy which has occasioned the ruin of the British deep sea fishery, both as a nursery for seamen and a source of national prosperity.

Patrick Morris, Esq., who from long residence is well informed on the subject, says in his Review of the state of the Colony of Newfoundland : " Con- " cessions of fishing rights to France and America, " and the effects of bounties given by these States " have rendered competition by British fishermen " impracticable, and have driven 400 sail of vessels " from the banks." In other parts of his review he shews that these vessels have been replaced by even a greater number from the ports of France— stimulated by a bounty of ten francs (or eight shillings) per quintal on the fish they bring home, and by an additional premium of five francs (or four shillings) on the re-exportation of fish to Portugal, Spain, and Italy, in all which markets the British fish is now undersold and consequently excluded. The premiums thus given are equal to the prime cost of cured fish as sold in the markets of St. John

and Halifax for exportation! Surely when the
concessions were made to France of the islands of
St. Pierre and Miguelon, and of all the northern
and part of the eastern shore of Newfoundland,
they should have been granted with such reser-
vations as would have prevented consequences so
injurious. The case of the Fishery on the banks of
Newfoundland is not one that ought to be considered simply, as of profit and loss, or otherwise
than as viewed by the French Government, who
annually disburse a large sum, said to be upwards
of 10,000,000 francs (or £400,000) in order to
stimulate a branch of commerce which places at
their disposal seamen sufficient to man more than a
dozen sail of the line. A fact, which, occurring at
a distance, is either unknown or discredited in England ; otherwise it is impossible that it should not
have excited the attention which its importance
demands.

Upwards of 200 American vessels fishing by
treaty in our waters, receive a bounty on tonnage
from the Government of the United States, whilst
the British fishery not only does not receive any encouragement whatsoever, but is actually burdened
with all the extravagant expenses of a regular
Colonial Government.

In Newfoundland, there is no income whatsoever
derived from the soil—no mines in activity—no
capitalist having interest from his funds—no foreign
trade or any trade at all, except in articles of mere

necessary consumption, with the exception of fish, unprofitably cured and excluded from a remunerative market, by reason not only of the premiums granted by rival Governments, but because the burthen of every shilling expended on unaccountably numerous and useless officials, falls exclusively on the club of the seal killer and on the hook of the fisherman—the amount whereof, annually collected, or obtained on loan, is nearly £80,000.

To this heavy exaction might be added the burthen arising from the extravagant expenditure incurred in the erection of private mansions, and public edifices, after the destruction of the town by fire, absorbing an income of upwards of £30,000 a year. I shall not now dwell on minor evils, but present them in the form of a statistical chart, equally applicable to the financial exactions throughout the West Indian Islands ; and conclude in the words of the Honourable Mr. Morris, Member of Council in Newfoundland :—

" The causes which have transferred the bank " and deep sea fishing, from British to French and " American vessels ought to be investigated, and a " remedy applied."—(*Morris's Review of the State of Newfoundland.*)

CAPE BRETON.

We remained at St. John's from the 8th to the 13th August, and then weighed with a view to visit Cape Pine, on which a lighthouse is about to be

erected. On our course along Newfoundland the
shore consisted of an uninterrupted range of barren
mountains, over which a thick fog having arisen
before we arrived at our destination, and such
weather being in this quarter generally of long
duration, the course was altered for Sydney in the
island of Cape Breton, where, instead of anchoring
off the Coal Mines as on our former visit, the Wellesley beat up to the town of Sydney, which is most
agreeably situated with well cultivated fields around,
altogether affording a pleasing contrast to any
maritime town in these Provinces. The harbour
is capacious and safe. It is astonishing that emigration has not been directed to this quarter—the
whole interior of the province on the Great Salt
Water Lake (or Mediterranean as it may be called)
is comparatively unsettled. The farms of the
Mining Company, on which there are crops as good
as any in England, prove of what the soil is capable when cultivated by those possessing knowledge
and capital, or their substitutes, industry and perseverance. The town contains many very respectable
inhabitants, whom I had the pleasure of seeing on
board at a little morning's dance that was given to
them.

The minerals here, like those at the Mines noticed
on our former visit, are all carboniferous, and the
schistous strata exhibit the most perfect impressions
of vegetable productions. The coal field in this
Province seems quite inexhaustible, and would of itself

have constituted a most valuable grant to His late
Royal Highness the Duke of York, without including all minerals and other substances, such as
limestone and clay, originating a very plausible
cause of complaint by the public at large, the whole
having devolved through the creditors of His Royal
Highness on the " Mining Company." However
obnoxious such extensive grants seem to be, it is
nevertheless undeniable that the important works
which they have undertaken could not have been
prosecuted either by the enterprise or capital of
the colonists themselves.

If I could conscientiously, I would willingly
avoid noticing the complaint of the Annexionists
(as those colonists are called who desire an union
with America), that a brick-chimney cannot be
attached to their log-houses, nor a wooden chimney
be lined with clay without liability to prosecution
for an infringement of the rights of the Mining
Company. This, as well as the greater grievance
of being prohibited from quarrying and burning
lime for agricultural purposes, ought to be redressed,
either by the Government at home, or by the renun-
ciation of the Company of rights so vexatious to the
colonists.

On leaving Sydney on the 28th of August, we
again ran down the eastern shore to Cape North,
but instead of pursuing the course to Newfoundland
we hauled up to the westward to circumnavigate
this island also.

The northern shore is precipitous, being constituted of lofty mountains, whose base is washed by the sea, thus exposing the tortuous and broken strata now again united by zigzag veins of quartz (resembling the flashes of forked lightning), bearing evidence of great electric action at the period of their elevation, and perhaps subsequently. Towards the north-west point of the island coal measures again appear, and coal is said to crop out a short distance inland, but it did not seem to be of any importance in a naval point of view, as the mines of Sydney afford an abundant supply.

The passage of the Gut of Causo presents picturesque scenery, and suggests considerations of an interesting character. It seems as if the island of Cape Breton had by some convulsion of nature separated from Nova Scotia. Massive hills of granite protrude through the coal measures, and raise them to a perpendicular position at Ship Harbour and in Coal Bay, and probably across the channel which separates the island of Arachat from that of Cape Breton, as the minerals on both sides are perfectly similar in their nature and position.

Having passed the Gut we anchored at Petit Passage on the 2nd September, whence I went in a boat to examine the practicability of connecting the Bras d'or (the noble lake before mentioned) with St. Peter's Bay, thereby to open a passage for the timber and produce of the interior, towards Halifax, as the advantage of water communication would greatly add to the value, and promote the settlement of the central part of the island. The distance necessary to be cut is not 600 yards, almost on a level; consequently it could be effected at a trifling cost. The object is of such importance as to be highly worthy the liberality of the parent Government. On the following day we proceeded towards Halifax.

Nova Scotia.

Persons generally know less of the vicinity where they reside, and of surrounding objects, than of more distant parts casually visited, and so it is in my case in regard to Halifax, save as to a general outline. In saying that the harbour is excellent, the town well situated, and its appearance picturesque on entering the port, is merely repeating facts already known; the surrounding country is sterile and unproductive, even as compared with other parts of Nova Scotia. This proceeds probably from the rocks being grauwacke, whose decomposition is slow, and when completed forms a soil not conducive to vegetation. When broken by violence, it splits into thick plates of a quadrangular form, containing much pyrites—it is curious to observe the perpendicular position of the strata of this grauwacke over the whole country, and that the ends of the fracture, wherever they have been protected from the weather, are invariably highly polished, as if by the scouring of enormous masses of detritus, the more weighty parts of which have indented scores uniformly in a N.W. and S.E. direction, all parallel to each other, and nearly transverse to the strata. The opinion offered that these ruts have been occasioned by stones fixed on icebergs is quite irreconcileable with obvious facts throughout extensive tracts, which I witnessed; nor is the belief that they were brought from the north-west, in my opinion, better founded. There is an enormous boulder, about three miles from Halifax, called the Rocking Boulder, often visited by the surrounding population from its vibrating, on a base of granite rock, at its point of contact. This boulder I thought might afford some clue to the direction whence boulders came, it being quite alone, and of great insistent weight. I therefore took some marines and a boat's crew to clear away the mossy soil, and uncover the rut (if such there was), by which it had arrived at its extraordinary position, on the apex of a mount of smooth and solid granite. A deep rut in the granite was presently uncovered on the S.E. side, which was traced from the stone into a contiguous lake, over which it had probably been forced. This rut was highly polished, and obviously not formed in the original rock, in which there was no flaw or indication of its having been excavated by any other cause than by the progress of the stone. We endeavoured to trace its route on the opposite shore of the lake, but our efforts were unsuccessful, the whole of the margin being covered with wood, the roots of which it would have required much time and labour to clear away.

Dissenting, as I do, from the opinion that icebergs transported the masses of boulders which are so frequently to be met with, and that they were brought from the N.W., it appears to me probable that the multitudes and mounds of boulders may have been torn off or transported by the effect of a comet passing from south to north, cumulating the waters which under the influence of rotation might constitute a spiral south-east flood, capable of producing all the phenomena observed, or by a gaseous comet from the south-east.

The whole south coast of Nova Scotia that I have seen consists of granite, and much is interspersed throughout the interior of the country. It shews itself in immediate contact with the grauwacke on edge, there being no perceptible intermission between these substances.

On Sambo Island, whereon the lighthouse is placed, the original granite, which is grey, has been vertically split, and the opening has been filled by granite of a darker colour than any I have seen in other quarters of Nova Scotia; it is strange that where all this volcanic power has been exerted there is no indication that I have seen or heard of, evincing that a crater, in the usual acceptation of that term, has any where existed.

I cannot say that I perceive much improvement in the city of Halifax since my first visit in 1794.

There are more houses, but few are in a better condition. Such limited portion of the country as I visited had not made the progress that might have been anticipated in one-tenth of that period. There seems to be no encouragement either to agriculture or commerce, and consequently there is a lack of exertion. A few establishments for rude manufactures were undertaken some years ago, but have all been given up in consequence of the one-sided free-trade with the United States, whose produce and manufactures undersell those of the British provinces, whilst articles originated in the Colonies are subject to heavy duties in the Union. As an instance of the strictness with which duties are there levied, the proprietor of a gypsum quarry, (a substance in great demand by agriculturists in the States), having ground that material in Nova Scotia, was charged with duty upon it as a manufactured article.

All are anxious about a railroad to Quebec, and I learn that by the exertions of the talented Colonial Secretary, a conditional sanction has been obtained to that measure if the provinces acquiesce in the terms proposed. The speediest means, however, to advance the interests of these provinces is to lighten the burden of taxation, which, though in appearance small, is in reality oppressive to the multitude who cannot afford to pay. The roads in Nova Scotia are few and bad ; and I do not believe they amount to as many miles as it is said there are Road Commissioners. Here, as in Newfoundland,

and indeed in each of the Colonies, there are officials sufficient to stock a kingdom. Let the expenditure of the public revenue be curtailed, and the taxes diminished on all articles to the indispensable minimum, and give a stimulus to agriculture and to the timber trade by encouraging the supply to our southern Colonies. But this, I may be told, is contrary to the principles of Free-trade,—there is no free-trade in the true acceptation of the term on this side of the Atlantic, nor can I learn that it exists in Europe. (See Appendix).

So long as our North American possessions shall continue attached to the British Crown, their interests ought not to be surrendered to the views taken of the presumed interests of Great Britain, but considered as a part of the whole, even if maintained at the cost of some slight sacrifice; but such policy would not demand a sacrifice, but on the contrary would promote the manufactures and commerce of the mother country—now in process of being superseded by those of another State. It is incumbent on the Legislature of England to investigate the causes which impede the progress of these valuable possessions.

Under existing circumstances it would be in vain to hope that the elected House of Assembly and the selected Council can remedy existing evils, as there is between them a struggle on almost every question—sometimes the one and sometimes the other pertinaciously advocating their favourite measures. It is unavailing to contend that the excellent Governor possesses a veto on measures of which he does not approve. He is powerless for good, and opposition on his part would only diminish his popularity and make matters worse. His Excellency Sir John Harvey has long been practised in the responsible situation of Governor, and I am persuaded, laments the inefficiency of the office he holds.

If I am asked what remedy is there, besides alleviation of fiscal exactions, for this paralysing state of affairs, I answer, Let the Crown land be *given* to respectable agricultural emigrants in portions proportioned to their means of cultivation, thus leaving in their pockets the amount requisite for clearing and cultivating the soil, nine-tenths of which, under the present system, lies unproductive in the hands of penniless and incompetent occupants.

That this remedy would be cheap also may be inferred from the fact that the sale of Crown land never has returned a revenue, beyond the salaries of the individuals selected to preside over that department, and who are in as complete possession of the proceeds as if all the Crown lands had exclusively been devised for their especial benefit.

BERMUDA.

My early impressions in relation to the Island of Bermuda, where I spent many happy days when youthful vigour made me more sensible of pleasur-

D

able scenes, are still vivid in my recollection. The beauty of its scenery has not changed, nor are its inhabitants less steady and judicious than in former days, but the medium through which those objects are viewed does not present them in such lively colours. I was in one of the first ships which entered within the reefs after the flag-ship of Admiral Murray had proved its practicability. Ships then did not proceed further than that anchorage, which obtained the appellation of "Murray's Anchorage," from the circumstance above mentioned. They now run in to Grassy Bay, 15 miles further within the reef.

The forts which crown the hills were not then in existence. St. George's, the capital, was a quiet village where few troops were seen, and the harbour was a resort of small colonial vessels, to the reception of which it was limited previous to the deepening of the channel. This has been effected at a considerable expense by the Legislature of the island in the hope to induce the large steam packets to avail themselves of that port.

This island, ever since the discovery of the opening in the reefs (by Captain Hurd in his minute survey), has been deemed of much naval importance, and plans were formed by the highest military authorities for its defence. A naval arsenal also has been designed for the accommodation of a large establishment of ships of war.

Distant islands, however, cannot be defended on

principles which would be the most judicious at home—by the erection of forts in all quarters that could be occupied by an enemy. It is obvious that under the circumstances of Bermuda, troops cannot be spared from the parent state permanently to garrison a multitude of forts, which on such a principle of defence would be requisite ;—if they could, the expense would be enormous, and therefore I cannot dismiss this subject without an expression of my satisfaction at intelligence I lately received that such extravagant and unavailing system of fortification has been suspended. In my opinion it is a great error to imagine that naval officers are unfit to be consulted respecting maritime defences—had it not been for so mistaken a notion, many hundreds of thousands of pounds, perhaps I might say a million, might have been saved. I unhesitatingly assert that gun-boats not only would suffice, but are by far the most available, and infinitely the cheapest defensive force amongst the rocks around the island of Bermuda.

The coloured population of this island are a fine race, incomparably superior to the generality of the coloured population in the West Indies. They are accustomed to navigate in their commercial vessels—their lives are almost spent in boats, and no better crews could be got for the defence of their own island than they would prove themselves to be. In this there would also be other advantages—the permanent pay of troops—the cost of barracks—

and above all, extensive stores, and reservoirs of *water* necessary for a large military force, would not be requisite.

The Naval Yard, if really planned by naval men, may indeed be adduced as a proof that I have fallen into error by imagining them capable of deciding on the best means of maritime defence. Nothing can be more inexpedient than all that has been done or is proposed to be done—blunders which, were I to set them forth, would seem incredible. The measures which have tardily been resolved on to stop the progress of an interminable series of fortifications, I hope may be followed by putting a stop to the extravagances perpetrated in the erections of this naval arsenal. It has proved a subject so vexatious to me that I shall not proceed in detail, but express my earnest hope that those who have the power will investigate the subject, and ascertain and pursue the course most useful to the naval service and least onerous to the country.

It is gratifying to turn from these military and naval incongruities to the state of the island, which is infinitely superior in all points of view to anything that is to be met with in the West Indies, or in the North American provinces. The Government of Bermuda has no debt—the legislature is economical in its disbursements, and has funds to appropriate to useful purposes, such as in deepening the channel into the harbour of St. George's, and in opening a channel into the Great Sound, a work which ought to have been undertaken at the public expense for the accommodation of the naval service, in preference to the formation of the Camber at the Naval Yard, which would have saved at least £150,000.

The Great Sound is one of the most magnificent basins of water that I have ever seen, and requires but the remains of a small barrier of rock to be removed, most of which (by the liberality of the Legislature) has already been thrown down at a cost of £1500, in order to open a harbour far superior to any in the United Kingdom, and competent to vie with all the ports throughout the world, in security, capacity, and the conveniences essential to a place of naval equipment.

The civil government is conducted in a manner highly creditable to the Governor and to his Council, who seem to have but one view—the prosperity of the island, in which considerable success attends their endeavours, in so far as they possess the necessary power. Horticulture, for it cannot be termed Agriculture, is progressing rapidly, and the island is being converted into a hot-bed for the production of early vegetables to supply the markets of the northern parts of the United States. Orange trees are again being reared, and as formerly their fruit may become a staple commodity. Nothing can exceed the richness and fertility of the vegetable soil which has been washed from the upper grounds into the hollows, most of which have hitherto been unreclaimed, by reason of their having no natural outlet for the water, and being nearly at the level of the sea.

On the climate of Bermuda nothing need be said; its mildness during the winter months is universally known, as well as its comparatively healthful temperature as contrasted with that of the islands in the tropics. No period of the winter is so cold as to prevent the growth of all European vegetables, nor is it destructive of tropical plants. It is remarkable that the common potatoes grown here during the winter have never been affected with rot, though those produced in the summer have been subject to that disease; the sweet potato wholly escapes.

The only drawback to the certainty of an ample return for horticultural labour is the violence of periodical winds, which in a few hours blight everything that is exposed to their destructive action. From this calamity, however, hedges of reeds, a species of cane, form a good protection.

The water at Bermuda which is derived from wells is strongly impregnated with calcareous and decomposed vegetable matters, and is neither wholesome, nor fit for washing. There is no spring in the island, although water is found everywhere at the level of the sea. This arises from rain being retained in the porous stone as in a dam, much of which percolates from the swampy hollows before alluded to. The water in the navy wells is of this description, and never fails to produce bowel complaints, especially to those unaccustomed to it. There are, however, naval tanks, containing uncontaminated rain water, five miles farther from the anchorage, and new tanks are being constructed in the naval yard, but these latter are within range of the sea spray, and no doubt will be attended with diarrhœtic consequences—and with this further disadvantage, that the expense of construction in that locality will be very considerable; whereas good rain water might be collected at a fourth, nay, at a tenth part of the cost, in a convenient situation, where the spray of the sea can have no influence.

I may here venture to assert, that were a Commission (whether scientific or not) of persons of good understanding and unflinching character, sent on a tour throughout these Colonies, to investigate and report on all matters of interest, the good they would effect would far exceed in degree any expense that could attend that judicious measure, and might lead to the adoption of salutary reforms and immense pecuniary savings. It may be said we have a sufficient number of Blue Books on grievances and reforms, and abundant information on Colonial subjects (which are neither read nor attended to), but I am persuaded that if the report of such unflinching Commissioners were to be published in the humble form of a pamphlet, it would be generally perused, and be productive of more beneficial consequences, than the reports in quarto of all the Commissioners who have been appointed within my remembrance.

The existence of this solitary island so far from the continent of North America is a circumstance meriting the attention of geologists, as well as the uniform material of which it is composed. It is all of a calcareous nature, but differing in condition from any of the other islands of the same substance. The strata are exposed in the perpendicular cliffs on the sea-shore in numerous precipices, from a hundred feet to minor altitudes, and are composed either of the most minute shells or parts of shells so triturated that they scarcely indicate their origin. In some places, however, there are laminæ containing shells in a more perfect state, all of a white colour, with the exception of one, (which I found on digging a cave) of a semicircular shape of a red colour, and almost as large as an oyster shell.

The whole of the substance of Bermuda can be burnt into good lime, but there is an indurated calcareous stone, often containing many perfect shells, on the island on which the naval yard is being built, which is preferred as more adhesive and better in quality.

Although there are no indications of volcanic products on this island, yet it exhibits manifest proofs that volcanic force has raised it from the depths of the ocean. In what stage of induration it was at that period, it is difficult to conjecture. The hills and vales throughout the whole extent of Bermuda have the stratified calcareous material generally conforming on all sides to the inclination of the surface. There are, however, many situations in which the strata present themselves as manifestly broken by force. In the deep cutting in the road which enters into the enclosure around the Government House, one of these breaks appears at the apex of the hill dividing its sides, which here incline towards the centre, exposing a wedge-formed supplementary part that fills up the interstice. In the grounds of the Admiralty House, curious instances of unconformable strata are laid bare in old quarries. These indicate some other cause for their nonconformity than that before assigned; and I am quite at a loss to imagine how the stratified materials could have been placed one above another at such different angles by the action of water, or in any other way, without appearance of disruption.

There are caves upon this island containing large stalactites. There is one on Tucker's Island where these stalactites reach from the top of the cave far below the surface of the salt water it contains. I am not aware of any other instance where similar crystalizations have taken place under the sea water. It seems to lead to the belief that this island was at some time less submerged.

There are other caves much larger, and one which goes in so far that the officers who accompanied me did not scramble to its end. This cave is formed by two large masses of calcareous matter having been reared up one against the other. I have seen some very beautiful crystalizations taken from another cave recently found in a quarry at Ireland Island, but the absence of petrifactions here (for I have never seen one), constitutes a remarkable difference between this formation and that on the island of Antigua, where the roads are almost made with petrifactions.

In clearing the surface of the rock, as has lately been done at the quarries, and in laying the foundation of the new convict barracks, the most irregular formation is exposed. Large holes are found contiguous to each other in the white calcareous rock, which are filled with a substance resembling chocolate in its colour, unlike anything else upon the island.

On the whole there is little to remark on the strata of Bermuda ; it might be curious, considering its surface as once the bottom of the sea, to put down a bore, and ascertain what is beneath ; for I do not believe that the calcareous portion exposed to view is of any considerable thickness. The effect of air and water produces a hard crust on its surface to the thickness of an inch, which prevents the unremitting action of the waves from encroaching with greater rapidity on its shores.

On the northern side of Bermuda, and at the distance of about twelve miles there is a reef, raised by the industry of the numerous species of coral insects, in from six to ten fathoms water and to the extent of 45 miles in length—exceeding in magnitude all the works of man for the protection of anchorages throughout the world. I have broken off some curious specimens of their work, but still more curious is the bottom in ten or twelve fathoms water, where forests of large submarine productions of innumerable kinds luxuriate like plants, exhibiting in appearance through the clear and pellucid water, varieties of form and colour such as are nowhere to be seen on the surface of the earth.

On the 26th January Her Majesty's ship Wellesley left Bermuda, in tow of the Scourge, the wind being light, and adverse to passing the Narrows. Having gained an offing, the Scourge was ordered to perform the like service to an emigrant ship, which a short time before had entered the Port of St. George's, destitute of provisions, having been at sea upwards of seventy days, and doubtless all must have perished had adverse winds detained the vessel a few days longer. A subscription relieved their immediate wants, and also provided for their proceeding to New York. The case of this vessel affords further evidence of the necessity of better supervision of emigrant ships at home. For the service of towing (which probably saved a delay of several days on their voyage) these people shewed their gratitude by cheering. It is a sad fate to be compelled to quit for ever friends and relatives, a cherished home and native land.

The Scourge being again sent a-head to tow the Wellesley, we averaged about six and a-half knots, until at the expiration of forty-eight hours a breeze sprang up from the N.E., whereby the Wellesley was enabled to return a similar service to the Scourge, which, like most steam vessels, is inadequately rigged, and thereby deprived of more than half the power of performing useful service. This defect is however in some degree remedied, when a sailing vessel is in company, for the Scourge advanced the Wellesley 220 miles, and during the next three days we towed that steam vessel 360 miles ; thus we alternately proceeded until our arrival at Carlisle Bay, Barbadoes, on Monday morning, the 5th February.

BARBADOES.

Here we found Her Majesty's ship Trincomalee, fourteen merchant ships, and as many brigs and smaller vessels. I went at 11 o'clock, under a broiling sun, to pay my respects to his Excellency Colonel Sir William Colebrooke, the Governor, and also to Lieutenant-General S. H. Berkeley, Commanding the Forces, both of whom reside at a short distance from Bridgetown, in houses admirably calculated for the climate. The reception room of each reaches in the form of a cross from side to side of the building, terminating under an open veranda. Four bed-rooms are, I believe, provided in the corners, by this arrangement.

On our way out we observed from fifty to sixty men, peculiarly dressed, working with an energy which I had rarely seen in any climate, and never within the tropics. On inquiring what they were (supposing them to be Coolies), I was informed that delinquents here were sentenced to a certain

amount of task-work, instead of a specified duration
of punishment—an admirable mode of administering
justice, which ought to be adopted generally
throughout the islands.

Windmills are extensively employed in this
island, and perform all mechanical duties required
on sugar estates; for this purpose nothing can be
better adapted by reason of the constant trade
wind during the day. The Governor has one
applied to pump water for his garden, which contains many truly beautiful flowering shrubs and
creepers. The well water is hard, being filtered
through calcareous substances; the water used for
drinking and culinary purposes is derived from
rain collected from the roofs of the buildings.

On the edges of the road leading to Government-
house, I observed wherever the strata were cut
through for the purpose of levelling the surface,
that the formation was of the same nature as at
Bermuda, though probably more recent. Coral
of considerable dimensions, large and almost perfect
shells, and numerous submarine productions are
cemented together in all the confusion in which
accident could place them. I saw no indication of
any other material on that side of Barbadoes,
though the cliffs on the shore are of considerable
altitude. I understand, however, that primary
rocks appear, and a bituminous spring exists therein
at the northern side of the island.

The conversations with the Governor and General

were brief, but not on indifferent subjects ; the mortality of troops in the barracks ; the freedom of detached working parties from fever, and their conjectures as to the cause, interested me greatly, especially as I learnt at the same time that the Medical staff had come to no definite conclusion.

During a restless night, it being calm and the thermometer at 84°, I determined to visit the barracks in the morning, and form my own opinion as to the cause of the great mortality amongst the troops. The able Deputy Inspector of the Wellesley, Dr. J. W. Johnston, officially accompanied me, and Captain Goldsmith, and Lieutenant the Hon. A. A. Cochrane, the Flag Lieutenant, were of the party.

It was not quite daylight when we left the ship to land at the Engineer's Wharf, close to the barracks, but scarcely had we approached within 500 yards of the shore, when a marshy exhalation charged with more than vegetable effluvia led to the suspicion that the barracks were placed in an unwholesome situation.

On landing, the most luxuriant trees and shrubs, and a very neat villa, were observed on the left. On looking in at the gate it was perceived to be barred and the villa abandoned, the inhabitants having been compelled to remove by reason, as I was informed, of the contiguity of two swamps to windward ; but, as my object was to visit the barracks, I did not personally verify this fact.

Turning to the right from these luxuriant trees and shrubs towards the barracks we walked first along the shore as far as the burial-ground, the whole of which is low, undrained, and overgrown with the most rank vegetation. We learnt that into this swampy ground various drains from the contiguous barracks terminated, and, of course, during rain evacuated their contents. This we could not verify by actual observation, by reason of the sharp-pointed plants called " Spanish bayonets" which cover the soil.

This low ground being to leeward of the barracks, although charged with pestilential exhalations, could not satisfactorily account for the fever; we therefore directed our attention to the barracks, and to an inspection of the ground to windward. On our way we noticed with surprise that the first range of barracks had their ends presented to the invariable course of the trade-wind, consequently the long verandas, and the numerous doors and windows leading into the apartments were supplied with air that had already been contaminated in apartments to windward. This position of the building was a great oversight in a climate where the thermometer stands this day (being mid-winter) at 88°.

As these were the barracks in which the fever commenced we might at once have concluded that their unwholesome position was an adequate cause of the disease; but as many soldiers had subsequently fallen ill in disconnected quarters, farther to windward, as well as officers whose quarters stood transversely to the tropical breeze, we pursued our investigation, and ascertained that a swamp had once existed on an extensive flat, now termed an exercising ground, although overgrown with rank grass knee deep, which is periodically cut for hay. Two large ricks proved the abundance of the previous harvest, and a pair of large scales on a lofty triangle denoted that it was applied to specific purposes.

This hollow is said to have been drained, and large sums must have been spent in the attempt to obtain that object, from the costly masonry used for that purpose. Nevertheless, the means employed have either been originally inadequate, or deposits have since blocked them up, as is manifest from the rank luxuriance of the harsh and resinous grass, whose roots rejoice in abundant moisture. This fact we personally ascertained by walking across the swamp, although warned by a large placard that privates or civilians were not to trespass.

I am convinced that this marshy meadow, called a Parade, but still in reality a swamp, is one great cause of the unhealthy state of the barracks at Barbadoes.

The Naval Hospital at Barbadoes, occupying a healthy situation, having been done away with, seamen were subsequently consigned to the Military Hospital, situated like the barracks, close to the swamp. The mortality in this hospital, as in the barracks, caused it to be abandoned, and the Captain of the Trincomalee had most judiciously sent several of the worst cases that occurred in that ship to the hospital at Jamaica, distant 1000 miles, for cure.

I unhesitatingly assert that troops can never be secure from fever in barracks so circumstanced in regard to swampy exhalations, excluded too as they are by a hill, or rather precipice, from the only wind that blows.

I may notice an additional cause of illness, namely, that the troops drink water collected from shingled roofs, which is always avoided at Bermuda, where nothing but rain water is used. Hence such water by reason of the alternate showers and sunshine, must be much more impregnated by the decomposition of wood and insects.

Black soldiers at present mount guard on the deserted barracks. Why blacks should not be exclusively employed, now that they are free and contented, I cannot conceive ; for certain it is that the European troops quickly become of far less physical strength than the coloured race. At any rate, European troops need not now be barracked in the skirts of towns in which every hut has a fountain of new rum, at a cost so low that enough to produce a fever can be procured for a penny.

Troops, if now required on these islands, ought to be barracked on rising grounds, where they can breathe pure air, which, with good water, contri-

E

bute more to health than the efforts of all the medical staff in tropical climates.

It is not my intention to dwell on ordinary subjects, yet I cannot omit to remark that the negroes who are much more numerous on this island than on any other in the West Indies, appear to be well-fed and cheery in their dispositions. They live in small wooden houses resting on clumps of wood or blocks of stone, a mode of construction which enables them, when tired of or displeased with their locality, to transport them elsewhere. I was told that a street of stone huts, constructed for their use, is almost abandoned by reason of the immobility of such residences. I consider this locomotive propensity a favourable trait in their character.

Behind the barracks we stopped at a hut on the rising ground, whereon the barracks ought to have been placed, and assuredly I never saw a more contented scene. There was a young negro and, I believe, his wife, together with an old woman, perhaps the grandmother of the child she fondled; we made inquiry as to their mode of living, and they shewed us green peas, seasoned with red pepper, ready to be cooked, yams, and cassada bread, as good as oatmeal cakes. These peas grow on large bushes, and vegetables of all kinds surround their hut.

When we descended from this elevation, which overlooks the barracks and anchorage, we passed a deep well sunk in the coralline formation, from which we caused a bucket of water to be drawn. It had no unpleasant taste, though from the calcareous nature of the strata it doubtless contained lime in solution. I should by no means recommend such water for the use of the troops, even as a substitute for the water now collected from the decayed wooden shingles of the barracks.

We returned on board fully assured that the position of the military barracks to leeward of the swampy ground, designated a parade, and of the hill just alluded to depriving them of the tropical breeze, were jointly the cause of the fever from which both the officers and soldiers had so severely suffered.

A deputation from the Literary Society, offering me the use of their library, and other respectable inhabitants came on board during the day. It was gratifying to hear their opinions regarding Colonel Reid, their late able and excellent Governor.

Having caused Carlisle Bay to be examined as to the facility of obtaining a supply of fresh water for a squadron of ships of war, the Master reported that a private individual possessed a very copious spring, which bursts from the strata at about 100 yards from the beach, and flows through, a fenced property into the sea; nevertheless this water is a monopoly, being sold at two shillings per ton, which is onerous to commercial vessels, and would amount to a considerable sum if circumstances rendered it necessary to obtain a supply for a fleet or squadron.

The two small tanks, called "the Navy Tanks,"

not being above four yards square, are not only
totally inadequate to the purpose, but they derive
their water from the wooden shingles of the Com-
missariat, and contain a fluid unwholesome and
offensive, both in taste and smell.

Having nothing to detain the Wellesley at Barbadoes, we weighed anchor, and made sail for
Tobago, and anchored in Man-of-War Bay. A
transport called the " Princess Royal" accompanied
us, and beat the Wellesley many miles on the
passage.

TOBAGO.

February 7*th.*—Man-of-War Bay is spacious
and secure, with the wind prevalent in the tropics.
The surrounding hills are partly cultivated, producing on their declivities towards the bay, sugar
canes, cocoa nuts, and plantains. Much more land
has formerly been reclaimed than is now in cultivation. There is a sugar establishment, an Overseer's
house, and some dozens of Negro huts, built of
materials totally different from those at Barbadoes—
their sides are of every kind and size, of canes and
sticks put together in the most rude and irregular
manner,—some of them indeed having tumbled
partly down, are. nevertheless inhabited in that
condition. The young children run about quite
naked, in which state several of them enjoyed the
luxury of pigs, by wallowing in a stagnant pool,
which a small rill of water had formed within the
beach.

The wages given to stout Negroes here, I was informed, were three bits, there being ten to a dollar. I learnt that a proprietor had lately arrived from England with the resolution to diminish the wages paid on his property, but on looking minutely into the state of affairs, he changed his opinion, and actually increased the amount. It is the custom here to set task work, but not to encourage more work to be done, which under the asserted scarcity of labourers, appears an omission, as more labour might be performed, and would assuredly be equivalent to an additional number of hands.

There is no good watering-place for ships in this bay, as a heavy surf rolls on the beach. The abrupt rocks which present themselves throughout a great portion of its circumference, are chiefly micaceous slate. The lofty mountains, and their cliffs washed by the sea, seemed to have been forced up by volcanic power, though no vestige of igneous products any where appears.

In running along the northern shore of Tobago, we observed that there is very little land cultivated, although there is the appearance, from brushwood supplanting trees in many places, that the land had formerly been cleared.

On proceeding along the shore, we do not perceive any cross section of the strata, by reason of their conformity to the elevation of the hills. The formation, however, wholly changes at Courland Bay, where a coralline and shelly rock, precisely similar to that at Barbadoes, has been upheaved. All this western portion of Tobago is low, sterile, uninhabited, and from what I learn, unexplored.

TRINIDAD.

February 8th.—We ran down towards Trinidad, the northern part of which island is formed of a range of mountains, some of considerable altitude, which exhibit a remarkable distinction from the northern shore of Tobago, by exposing the stratification in the precipitous rocks, washed by the sea, these being perpendicular to the elevation of their sides. I am impressed with an opinion, that valuable ores exist in the mountains of Trinidad, in situations almost inaccessible, and I believe, wholly unexplored. The tortuous strata, which constitute the islands that form the Dragon's Mouth, show that violent volcanic power produced their elevation. I propose to take another opportunity to examine these islands and the northern shore. The night had set in before we reached the anchorage of Port of Spain, so that nothing was seen of the town until the following morning.

February 9th.—At 10 o'clock I went to pay my respects to his Excellency Lord Harris, the Governor, who resides about a mile distant from the town, in a mansion standing on a plain, too low for my taste in so warm a climate. There are many large and ornamental fruit trees, nutmeg trees, and shrubs of all descriptions, in the surrounding grounds, which are termed a Botanical Garden. The collection is fine, and is superintended by a botanist and gardener. Excellent water is brought in pipes from the lofty mountains in the rear, and several fountains give a cool appearance to the vistas seen from the spacious and well ventilated reception room of this mansion. His Excellency invited me to take up my quarters there during my stay, which I declined—having learnt that excellent timber for ship-building was to be procured in Trinidad, which I was anxious to ascertain the practicability of embarking in the bays on the northern shore, particularly in that of Maraccas, at the foot of a lofty mountain covered with large timber.

I proceeded on the following day in the Scourge, accompanied by my Flag Lieutenant, Mr. Cochrane, who is zealous and active in aiding in whatever can conduce to the interests of the service.

The anchorage in Maraccas Bay is good, and the practicability of embarking timber quite evident; but the impassable nature of the woods, occasioned by creepers entwined in all directions, rendered every attempt to mount the acclivities abortive. It is quite obvious, however, that were a pathway cleared, timber might be felled and hurled down, as in Switzerland and other places. The appearance of great volcanic violence is especially evinced by all the strata exposed to view around this bay.

On the morning of the 11th February we weighed,

and returned through the Dragon's Mouth, shaping our course for the great natural curiosity of Trinidad—the Pitch Lake—which I hoped might be rendered useful as fuel for our steam ships—so important in the event of war—as fuel is only obtained at present from Europe. The United States and Nova Scotia are never resorted to; hence, could this Pitch be rendered applicable as fuel—our vessels could be supplied when an enemy would be almost deprived of the use of steam in these seas.

We arrived at La Brea, and before daybreak on the following morning we were on the road to the Lake, or rather on a stream of bitumen (now indurated), which in former ages overflowed the lake. Indeed the bitumen beneath this road seems still to be on the move, as shewn by curvilineal ridges on its surface, like waves receding from a stone thrown into water.

The appearance of the Lake is most extraordinary—one vast sheet of bitumen extends until lost amidst luxuriant vegetation. Its circumference is full three miles, exclusive of the creeks which double the extent. The bituminous surface is of a dark-brown waxy consistence, except in one or two places where the fluid still exudes; obviously the spring is in full vigour beneath, for the whole surface of the lake is formed into protuberances like the segments of globe pressed together, having hollows between filled with rain-water, which (except in the immediate vicinity of the

bituminous springs) is inodorous and without taste—
an extraordinary fact—shewing that this bitumen
is of a nature quite different from that of pyro-
technic, mineral, or vegetable tar. In its dry state
it is quite insoluble in water, though when charged
with essential oil, as it exudes from nature's labora-
tory, it imparts a pungent and unpleasant taste.

A considerable quantity of gas bubbles up through
these bituminous springs, shewing that decomposition is still active amongst the materials whence it
exudes. Some of the recent bitumen has an odour
resembling vegetable gum. Mr. Johnson, the very
obliging proprietor of a neighbouring estate, had
the goodness to cause some of his labourers and a
cart to bring samples to the beach. Means of
transport, however, were so inadequate, that we
had recourse to digging the more impure pitch on
the beach in order to prosecute our trials for its
substitution as fuel. This bitumen, which had flowed
upwards of a mile from the lake, was combined
with earthy and other substances which it had
encountered in its course.

Various attempts have heretofore been made to
apply the bitumen to useful purposes, but without
success, as we may judge from the total abandonment of those trials and expectations which for a
brief period induced its shipment to England with
a view to its application to the pavements of London
and other cities. All excavation has consequently
ceased, and so low is the estimation in which the

bitumen is held that the duty on embarkation is only one halfpenny per ton.

The nature of this bitumen is very different from that of coal. When exposed to a naked fire it becomes fluid, and runs through the bars before gas is disengaged, or at least before it is raised to a temperature at which it will ignite; perhaps it requires more, or purer air than enters through the bars of steam-boat furnaces; a conjecture which seems to be confirmed by the dense smoke speedily produced.

Reflecting on these circumstances, it occurred to me, that were the inner bars of the furnaces of the Scourge withdrawn, and a quantity of ashes heaped up until the ash-pits beneath the removed bars were nearly filled, the bitumen thrown on would be absorbed as it melted, and would be gradually given out and converted into flame and heat by a current of air permitted to enter between the fire and the bitumen so absorbed by the heated ashes.

February 17*th.*—Having made this arrangement in eight of the twelve fire-places, the Scourge weighed and proceeded under steam towards the Hill of Naparima. During this short trip the bitumen was used in the furnaces of the Scourge in the proportion of 2 to 1 of coal, and produced an effect on the engines nearly equal to that which double *that* quantity of coal would give—that is, 2 parts of bitumen, under the foregoing arrangement, were equal to 1 of coal; 8 measures of coal, considered to be cwts., and 16 measures of bitumen were used per hour, and produced on an average seven revolutions of the paddle-wheels per minute, and $4\frac{1}{2}$ knots per hour, without sail. It must be remarked, that only eight out of the twelve furnaces were used, and that the amount of fuel consumed was little more than half the quantity appropriated to such engines.

At Naparima we learnt that coal was to be found at the village of Petit Bourgh—on examination, however, it was ascertained to be an exudation of the same nature as that at the Lake, combined with more earthy and ligneous matters. This compost burns without deliquescing and with less smoke, but produces little heat. Near this place there are two hot springs, one of which, we were informed, is of a temperature too great to be used in baths without being cooled. The fact not being interesting in a naval point of view, I did not delay the Scourge to ascertain it.

February 18*th.*—Lieutenant Cochrane brought off some specimens of the mineral strata from the cliffs on the beach, some of which were red, seemingly caused by the action of heat; others of a schistous nature were impregnated with bitumen, probably by the same cause. Mr. Cochrane again went on shore and brought off other specimens of fuel, and of the strata exposed in the cliffs on the shore.

We left Naparima satisfied that there was no

coal, nor any minerals which usually indicate its presence, and proceeded again to try the bitumen of La Brea by running about the Gulf of Paria, and ultimately towards the anchorage of Port of Spain. All this day a better result was obtained, and there is no exaggeration in asserting that even in the rude manner in which the furnaces were hastily fitted, the bitumen, measure for measure, produced one half the effect of the best coal.

A defect was obvious in this arrangement—namely, that the air to ignite the gases evolved from the pitch was interposed between the hot current from the coal fire and the ashes in which the pitch was contained, and consequently cooled down the gases below the proper temperature, which it required considerable heat to restore. I wished, therefore, to make a channel to carry the air necessary for the ignition of the gases of the pitch behind the ashes near to the bridge, but I had no opportunity of effecting the object.

Having informed His Excellency Lord Harris of the result of the experiments made on the substitution of pitch for coal on board the Scourge, His Excellency, who not only takes a deep interest in promoting the prosperity of the island under his authority, but in furthering the advancement of whatever can prove generally useful, was pleased to accept my invitation to take a tour of the island, and witness himself the effect of the combination of bitumen and coal for the use of steam vessels.

His Lordship embarked in the Scourge on the morning of the 23rd February, and proceeded along the northern shore of the island, viewing all the precipitous rocks, which evince strong indications of mineral products. We anchored at 8 P.M. in Maraccas Bay, which I had before visited with a view to obtain timber for the Naval service. On the following morning we weighed and again coasted along the island for the like purpose, and with the same impressions as we had received yesterday—namely, that the mountains of Trinidad, and I might add the plains, well merit examination by a mineralogist, as no scientific survey has hitherto taken place. Having rounded the N.E. Cape, we anchored off Point Matura, and on the following morning proceeded along the eastern shore, on which the surf, occasioned by the tropical winds, rolls in a manner that would render it dangerous to land from boats; we therefore continued our course, although desirous to have made inquiry respecting coal, which is said to be found near the Point of Manzanilla. My regret at being prevented from making such inquiry at this time was the less, because, if coal there be, the period has not yet arrived when moles or jetties could be undertaken to facilitate its embarkation.

On the eastern side of Trinidad three extensive alluvial plains are separated by two ranges of hills, less elevated than those along the northern and southern shores. These are covered with forests

said to be of very fine timber, and there are rivers
whereby, no doubt, it could be conveniently floated
to the beach, whence it might be shipped without
having recourse to the means requisite for the embarkation of coal. It is curious to observe the
different inclinations of the strata, which on the
northern mountains dip to the south—at the next
Cape to the north—the following again to the
south, and the southernmost reversed. There is
a note on the map of Trinidad recently published,
that the mountains on the S.E. part of this island
are inaccessible and covered with indestructible
wood. My curiosity was excited, but I had no
means of gratifying my wish to disprove the cor-
rectness of the former assertion, or render a service
to Her Majesty's Navy by a confirmation of the
latter. An unpleasant feeling was experienced by
the reflection that many valuable treasures, from
parsimony, or rather the small influence of scientific
men, may be unknown and neglected, whilst so
many and such heavy salaries are paid to persons
whose duties are merely nominal or of little utility.

We again proceeded on our tour on the following
morning, and effected the circuit of the island as far
as the anchorage at La Brea, within a mile of the
great Pitch Lake, to which His Excellency accompanied me, expressing his sincere wishes that
the attempt which he had witnessed to render the
bitumen useful for steam vessels might be accom-
plished, and also that the suggestions regarding its

applicability as a manure for the growth of sugar
on the worn-out estates of the long-cultivated West
India islands might prove well founded, and also
its use as a preventive of the potato disease and
turnip blight.

The seamen and marines of the Scourge, and
some blacks who were hired, were employed for
some hours in quarrying and embarking a farther
supply of pitch, and on the same evening we returned
to the Port of Spain.

Geologists may be curious to learn if any evidence has presented itself of the probable source
whence the bitumen of Trinidad has originated;
and although a decisive judgment cannot be formed
from the fact I am about to mention (witnessed on
my last visit to that island), yet it seems to be in
confirmation of the generally received opinion, that
the bitumen is the expressed juice of forests of
timber, amassed and fermented under circumstances
which the mind has difficulty to conceive.

On a precipitous bank of clay, washed by the
sea, the top of which is covered by exuded bitumen
not emanating from the lake, I found on cutting
down the precipice a horizontal bed, four feet in
thickness, resembling a seam of coal, perfectly
defined in its limits, though formed of all kinds of
branches and leaves. I have got specimens of this
material which was thirty feet under the bitumen.
Being anxious to ascertain the existence of the
mass of lignite which I inferred must be at a

greater depth, I commenced to bore, but by reason of the tenacity of the clay, the rods did not sufficiently clear their passage, and were finally arrested at the depth of sixty feet under the bed of ligneous substance before mentioned.

I see a boring rod advertised in the papers, which frees itself by the introduction of a stream of water. This no doubt would suffice not only to ascertain the curious fact above alluded to, but probably to put the British Government (or those who should undertake the enterprise under its sanction), in possession of the most abundant mass of anthracite in the world—a substance which might be supplied to our steam vessels in that quarter at less than one-fourth of the cost of coal, thus securing for purposes of defence and commerce the exclusive Steam Navigation of these seas during a period of war.

The plains of Trinidad have a fertile soil, which simply by clearing the ground is capable of being rendered the most productive in the West India Islands for the growth of sugar, and whatever can be cultivated in a climate the most uniform in its temperature—the most congenial to tropical plants —free from the evils of hurricanes, and from all impediments to vegetation. I am confident, that if the hands of his Excellency were not bound by restrictions and routine, the progress of Trinidad would soon verify this opinion. Lord Harris nobly tendered a portion of his official income in allevia-

tion of the burthens which are so severely felt in
the present depressed state of agriculture and commerce; but from some cause, his Lordship's liberal
intention was not realised. The example would
have proved salutary, as it must have been followed
by reductions throughout other West India Islands,
whose resources are even in a worse state than
those of Trinidad.

It has often astonished me to hear the salaries of
officials denominated vested rights. Is it reasonable,
that whilst the ground has ceased to be cultivated,
because production is unprofitable, not only the land
should continue to be taxed at the rate it was in
prosperous times, but that a duty should be levied
on the exportation of its produce? Is it reasonable,
that whilst householders can obtain no rent, and
have no income, save the bare means of providing a
scanty subsistence, they should be assessed at the
rack-rent of former valuation? Can any property
be more entitled to protection than that of the
owners of the soil, or of the dwellings they inhabit ;
and yet all these, as appears by the numerous
gazetted sales, are sacrificed to the collection of
sums, the bulk of which is uselessly and prejudicially
expended. (See Appendix.)

Whilst the Government of the Parent State has
alleviated the burdens on the productive classes, is it
just, that taxes on food, and on all the necessaries of
life, should be continued throughout the Colonies,
and that even their productions should be intolerably

F

burdened with local imposts, whilst complaints are loud and true of the absence of all remuneration from the sources which once constituted the prosperity of those now impoverished and oppressed possessions. The above observations do not apply exclusively to Trinidad, but to the whole of the islands, which scarcely differ in degree in the causes of ruin which seem irremediable by any authority except the Legislature of the Parent State. I am persuaded that the Chief of the Colonial Department at home would endeavour to counteract the causes of widely spread and increasing ruin, were he in possession of correct information; but popular representations of grievances, often embodying misapprehensions as to their true origin, and accompanied by suggestions of impracticable remedies, are denied or disputed in counter-statements by interested officials, so that the Colonial Minister is bewildered, and can form no correct judgment from such conflicting statements.

I hold it to be impossible, that the monstrous absurdities, and violations of every principle of good government which exist throughout these Western Colonies, could be tolerated an instant were their consequences known and believed by those in power, or were they laid before the British public by any person on whose judgment and opinion they could rely. Can it be credited, that even in the Island of Trinidad, not only multitudes of valuable properties are brought to sale from the inability of their owners to pay the fiscal demands, but that properties are consigned to the Government auctioneer even for so small an assessment as three-fourths of a dollar, which is nevertheless the fact, as appears by advertisements of sale in the Gazette of Trinidad? (See Appendix.)

The emancipation of the Slaves was a glorious act, but the rescue of these noble possessions from ruin, and the restoration of prosperity to an integral part of the Empire, would redound to the honour of whoever should successfully advocate the cause of reason and justice—not only on the principles of equity, but with the less noble view of gain to the Parent State, as it is certain that the consumption of British manufactured articles has fallen off in these Colonies to an extent which has not been counterbalanced by the increase of exports anticipated from the questionable policy of concession to Brazil, in which I have reason to believe the supply of articles required for the Slave Trade has constituted a large proportion.

On the 1st of March we weighed from Port of Spain to visit the other British West India Islands, and passed through the third channel formed by the Islands in the Boca del Drago. We had a near view of the tortuous and broken strata on both sides and of the cracks and openings, now filled with white spar. The average inclination of the strata is towards the south, as is the case with the mountain range on the north side of Trinidad. Micaceous schist and alum are the chief components of these precipitous islands.

GRENADA.

1st March.—Grenada now in view presents itself as a range of lofty mountains. On nearer approach they are obviously volcanic. The town of St. George's is situated on the side of a hill composed of alternate tortuous strata of scoria, lava, pozzolanna, and the products of igneous decomposition. The neighbourhood exhibits a magnificent landscape, such as none but a most able artist could delineate. Indeed, it is one of the most beautiful localities I have ever beheld. We anchored in the roads, but there is a harbour close by of considerable size and good depth of water, the entrance to which is quite protected by the S.E. promontory of the island. On the road ascending to the Governor's residence we had a most beautiful bird's-eye view of the town and bay. Here again were towering trees, and shrubs of species which I had never before seen, beautiful in colours and of unusual forms. I procured the seeds of some of these in order to plant them at Bermuda.

Wherever the cuttings to form the precipitous road laid bare the strata, dust, ashes, large and small pumice-stones, present themselves, all indicating the convulsions to which this island, in ages past, must have been subject. There is no appearance of active volcanic power at present, but there seems to me to be no security that eruptions like that at St. Vincent in 1812 may not take place.

The town of St. George's has seen better days. The houses, with few exceptions, are in a dilapidated state—paint once has been in use, but there is no indication of any having been applied for many years.

Numerous negroes loitered everywhere about the beach, but few were employed. The heat was so excessive that I had no means of personally doing more than take a cursory view, unless at the risk of indisposition. This day, the 2nd of March, the thermometer stands in the shade at 89°. The barracks are situated on the top of a hill, near to which appears the foundation of a Military Hospital. £14,000 was voted for its erection, but this sum, I was informed, had only laid a few courses of stone and defrayed the expense of iron frames sent from England, which I saw lying near the Custom House!

On quitting this anchorage we had an opportunity, by running along shore, to observe how little land is now cultivated when compared to that which bore obvious marks of having formerly been under the planter's hoe. I feel assured that at least three-fourths of the estates are now abandoned. I procured a memorandum of imports and exports, which establishes the fact of the great deterioration of agriculture and commerce.

It being calm and the Wellesley in tow of the

Scourge, permitted a very near view to be taken of Grenada, more general and equally correct as if each estate had been actually visited. To economize fuel two fires only instead of three were lighted in each boiler, which number evolved more heat than could be absorbed on the Admiralty principle of boiler-making. This arrangement is greatly preferable to using all the fires in a less number of boilers. We passed the southernmost of the Grenadines before dark. They appear all to be volcanic There are two conspicuous conical mountains, but the bulk of them seems to have been forced up from below in the state in which they now appear.

ST. VINCENT.

On the 3rd March we anchored in Kingstown Bay, opposite to the town, above which there are some good cane fields, and a little bustle on the beach, where an American vessel was landing a cargo of deals. There were half a dozen of vessels at anchor awaiting cargoes of sugar. It is said that some new process has enhanced the value of that produced here, which will prove a blessing if it shall extend to the islands we have already visited.

As at Grenada, all the strata denote volcanic influence. It is quite curious to see the manner in which ashes, stones, and lava, are intermixed. Here there is a vast mass of pozzolanna in precipitous cliffs on the shore, whence it can be readily shipped. It appears to have been formerly embarked as ballast, and probably thrown overboard when it had served its purpose. It is only a year or two since Mr. Anderson, the clerk of the works at Bermuda, endeavoured to ascertain whether pozzolanna existed in the volcanic islands in the West Indies, which, on inquiry from those supposed to have been most acquainted with the mineralogy of that quarter, appeared to have been unknown; he, therefore, obtained samples from the cliffs, which proved to be the identical substance hitherto procured from Italy, by which a great saving has been effected in the submarine works at Bermuda.

On the north side of the town I observed a limekiln, and on inquiry whence the limestone came, I learnt that it was brought from one of the Grenadines, which is of a coralline formation.

The island of St. Vincent, like Grenada, seems in regard to cultivation to be fast returning to its primitive state of brushwood and wilderness. I have procured, and shall, when possible, continue to procure from the cliffs on the shores of the islands we pass by, such specimens of strata as can be had without delaying the Wellesley.

On this island there is an enormous volcanic mountain, which, in 1812, burst forth with such tremendous violence that the dust and ashes were projected through the tropical breeze into the counter current in the higher region of the atmosphere, and were carried, thereby, to windward, and fell upon Barbadoes in such dense clouds as to obscure the light of day. This crater is at the present time said to be a pool of water 700 feet below the apex of the mountain, within a cavity so precipitous and awfully grand that few persons can trust themselves to look down into the abyss, which can only be viewed from an overhanging rock difficult of access.

The brief notice of exports which I obtained from the Custom House here shews the comparative decay of every branch of commerce.

The Lieutenant-Governor being absent on a tour in the interior, I did not go to Government House, which is on an elevated situation. I was glad to see that the Barracks and Military Hospital were on the top of what in other places might be called a mountain, free from marshes and enjoying uncontaminated the tropical breeze.

St. Lucia.

The wind being very light, recourse was again had to the Scourge to tow the Wellesley to Port Castries, or rather to the bay off that port, which is unsuited to a large ship, unless in case of necessity. This island has two most remarkably conical-shaped mountains at its southern end, each of which seems split in half—one part probably being blown away by volcanic action—like the island of Stromboli in the Mediterranean. The setting sun clearly shewed us the nature of the strata along the S.W. shore,

which exhibited the same appearances as at Grenada
and St. Vincent. Here the smouldering volcanic
fires emit sulphureous fumes, but circumstances did
not permit their being visited. The island is covered
with trees and dense brushwood, scarcely any part
is cultivated ; the air is impregnated with decomposed vegetable matters, increased by pestilential
exhalations from numerous swamps. The Governor
has just lost his wife by fever. From a cursory
view of this island, I should say it is pre-eminently
unhealthy, and that here more than on any of the
other islands, the practice of selling Crown land at
a high price, in its thickly-wooded and uncultivated
state, and afterwards laying a tax on its cultivation,
is most unwise and injurious in every point of view.

The barracks for troops, contiguous to the town,
are elevated, but I judge they are unhealthy, as
other barracks are constructed on a detached rock,
remote from any population, and protruding into
the sea from the N.W. part of the island, to which
it is connected only by a reef, over which the sea
breaks with great violence. I did not, however,
inquire respecting the health of the troops, under
the circumstances of the Governor's bereavement,
and it being Sunday, I had no means of informing
myself from any other quarter.

The day was extremely hot, and the rain fell in
such torrents on the mountains, that we were de-
terred from prolonging our stay on shore. The
town is situated on a flat alluvial soil, brought down

a ravine by torrents from the mountains, which rise abruptly from the shore. There is a small stream to the southward of the town, the sides of which are marshy. On this the Ordnance depôt is built, and close to that there is a burial-ground, seemingly as a warning of the consequence of residing there. Nevertheless numerous small residences are thickly scattered, in only one of which I saw a white inhabitant. There are about 19,000 blacks and 1000 whites at St. Lucia.

As soon as we returned on board the Wellesley, the anchor was weighed, and we proceeded again in tow of the Scourge, close along shore. The whole island seems one unreclaimed wilderness, scarcely a patch of cultivation being seen on its western side. Surely a thousand whites, and 19,000 blacks, if the lands were not so locked up and heavily burdened, might cultivate some part of this desert! This evil, common to all the islands, seems particularly injurious here, and I may venture to predict, that never, whilst the world lasts, will any man, having £100 in his pocket, invest that sum in the purchase of 50 acres of unreclaimed land at St. Lucia.

MARTINIQUE.

Having passed the northern extremity of St. Lucia, the course was directed towards Martinique, where we arrived at daylight on the 5th March. The S.E. part is low, but that to which we were

peoceeding presented lofty mountains of a conical
form, some of them having their tops cut off like
the mountains of Grenada, St. Vincent, and St.
Lucia.

We passed close to the famous Diamond Rock.

When the sun rose, the igneous origin of the
strata of Martinique was obvious in the promontories
washed by the sea. The same tortuous stratification
and intermixture of stones, dust, and ashes, clearly
shewed that they were of the identical nature as
the specimens we had procured in the other islands.
I wished much to have entered the Bay of Fort
Royal, to have learnt the state of the island, in
which the negroes had just emancipated themselves, but as the squadron was awaiting my arrival
at Jamaica, I did not gratify my desire.

We had a full view of the town and fortifications
of St. Pierre, under which a French frigate dis-
played her ensign, and also a steam ship and two
small vessels of war. The course was continued
close along shore, and a more striking difference
can scarcely be imagined than that which we
witnessed in the aspect of Martinique, as compared
to the British Islands. Here every spot is cleared
that can be rendered useful to man. Continuous
fields of sugar cane cover the whole surface of the
island, each of which is surrounded by neat hedges
of coffee-trees. So minute are the divisions in
precipitous places, that their appearance resembles
that of Madeira. Whether all the toil bestowed on

this island will henceforth become unavailing, it is impossible to foresee; but it will be lamentable if such shall be the result.

In looking through my best telescope, within a short distance of the shore, I did not in the whole extent of thirty miles, see a single coloured individual employed in rural labour. Numbers however flocked out of their huts to witness the novel sight of two large vessels running along without sails. Many boats were engaged in the easy task of fishing, and numerous nets were suspended on poles on the beach, which probably belonged to the owners of the contiguous estates.

I looked most attentively for smoke from the chimneys of the steam engines and boiling houses, and for cattle or poultry in the spacious yards attached to the proprietor's houses, but in vain—all seemed as if they were abandoned. The ripe canes, which covered the lands, and which ought to have been cut a month ago, proved that the emancipated population had not resumed their labours. Rear-Admiral Bruat (the hero of Tahiti), is at present Governor of Martinique, and no doubt has a more difficult task on hand than in conquering the people and subjugating the Queen of the aforesaid unfortunate island.

DOMINICA.

On passing the N.W. point of Martinique, we perceived in the distance two conical mountains of great height on Dominica. When we arrived within a mile of its shore, all the strata that presented itself to view consisted of volcanic matters; numerous conical hills, which had not the appearance of volcanoes, were formed of immense loose stones and ashes (perhaps pozzolanna); all the island seems to consist of igneous products. We went close to the town of Roseau, but as the soundings are deep to within two cable lengths of the beach, and the wind was on shore, a communication was sent to the Governor to express regret at not being able to anchor the Wellesley under such circumstances. We then stood on to Prince Rupert's Bay, a more eligible anchorage, at the N.W. point of the island. On the following morning we visited the fort and barracks, now almost in a state of ruin. On landing we met a very intelligent gentleman, who politely offered us horses, which we declined in order to be able to converse more freely. The view from the fort is picturesque, but our attention was chiefly directed to an inquiry into the prospects of the planters and the state of the negro population.

From this gentleman we learnt that the negroes had long stood out for high wages, under the impression that the profits of the planters were sufficient to afford them; but now most of them had returned to work, the proprietors having come to an arrangement to give half the canes to the cultivators, a judicious measure, in my opinion, as Dominica appears by far the best looking of the British islands,

which we have yet seen. There are vast numbers
of coffee trees, and the ground is said to be admirably calculated for the cultivation of that shrub.
Several new clearances were in progress, which
looks as if those already worked were doing well.

I must not omit to notice the fort and batteries in
Prince Rupert's Bay. There are about a dozen of
guns and carronades, mounted half way up the hill,
all in a very inefficient condition, some propped
horizontally, by bits of stick, and others resting on
rotten beds of timber. On the ground near the
enclosure there are fourteen 24-pounders, and half
as many smaller guns, dismounted, exposed to the
daily tropical showers, and the decomposing influence of the sun. One of these was lately brought
from England to Roseau, and thence to this bay,
where assuredly there are more guns than are
wanted. I hesitate to add the account I received,
that it cost £180 to place it in the situation where
it now lies. Seeing no magazine, I inquired where
it was, and learnt that there were 1300 barrels of
powder in the casement pointed out; 800 of which
had been there since the war, and on my asking if
care was taken that they were kept dry, and regularly turned, I was told, that about a-year ago,
when 400 additional barrels arrived, the previous
store had been shifted.

It is curious and also most grievous to any one
who has the interests of his country at heart, to
witness everywhere the wreck of valuable property

and the waste of public money. Thrown down from this fort, and involved in the shingle of the beach, lie the excellent French artillery and shot for which English guns were substituted at an enormous cost.

I was told that some estates in the neighbourhood of Prince Rupert's Bay were doing very well, and, on inquiring the particulars of this phenomenon, I was informed that money had been lent by a neighbouring attorney to the former proprietors; that he had foreclosed the mortgages, and thus had got possession of them for a trifle. The complete ruin of former possessors is thus proceeding with rapidity in all quarters.

March 6th.—On leaving this anchorage we passed over the waters whereon the Naval action of the 12th April, 1782, was fought, and I have taken the opportunity of refreshing my memory of the circumstances by reading the account given in Clarke's Naval Tactics. I shall not notice his criticisms of the manœuvres, or his reflections on the result. It is easy to perceive how actions might have been more successful, after their termination, but Rodney's breaking through the line with his own ship (the first time that manœuvre had ever been attempted) was a glorious act. I cannot so term the conduct of the then Administration, who superseded Rodney in his command on the very day on which, having repaired his shattered ships, he proposed to sail from Jamaica in pursuit of the remainder of the French fleet, which had gone to the United States.

The barren, rocky islands, called the Saintes, are in sight to the N.W., and Mariegalante to the N.E. of the scene of this memorable battle.

MARIEGALANTE.

March 6th.—On approaching Mariegalante its appearance is wholly different from any of the islands we have lately passed. It is flat and low, and seems, like Barbadoes and Bermuda, to be of coralline formation. It is highly cultivated, but I conclude that nothing is now doing by the lately emancipated slave population, as I counted seventeen windmills, not one of which was in motion, although there was a fine breeze; nor was there a chimney smoking, though it was the season for crushing and boiling the cane.

PETIT TERRE.

March 7th.—The small flat island called " Petit Terre " is also a coralline formation, so is the whole of the eastern side of Guadaloupe, as far as can be distinguished by an excellent telescope, employed to examine the sections of the strata formed by the action of the sea along shore.

DESEADA.

The island of Deseada, which, probably, was named when Columbus was so fortunate as first to see the land, is high ; the inferior portion appears to be of volcanic origin, but there is a horizontal white stratum along the flat upper surface, brought up from the calcareous bottom of the sea.

ANTIGUA.

March 8th.—From Deseada we shaped our course for Antigua, where we anchored in the outer bay by reason of there being only $23\frac{1}{2}$ feet water on the bar, and the Wellesley drew $22\frac{1}{2}$. The swell occasioned a lift of at least two feet, so that the ship would undoubtedly have touched the bar had we attempted to enter. It is said that this bar is formed by a narrow shelf of rock, which might be easily removed.

The buildings of the Naval Yard have lately suffered by the double effects of an earthquake and a hurricane, to both of whch evils this island is periodically subject. The massive sea wall to the northward of the Yard has sunk in several places. The store-houses and residences have been shaken and partly thrown down. As a rendezvous for ships of war assuredly it would be a waste of public money to repair the damages, as it by no means merits the fame it enjoyed in former times. It might, however, constitute a good rendezvous for the packets, which can here take in coal with facility.

The troops stationed in the Yard are black ; the white troops are kept on the hilly ground, where there are several Forts and Barracks. I inquired as to their sanitary state, and learnt that it was satisfactory.

Having inspected the remains of the Naval Yard, and transacted some uninteresting official business, I walked out to an isthmus dividing the harbour from a spacious bay. On the road-side we observed a quarry, in which there was a stratum of broken-up stones, involved in matter that had been in igneous fusion. One stratum is a coarse granite, which is very hard, and has been employed on the wharves of the Dock Yard. I inquired where the lime was procured, and was informed there was abundance of limestone on the island.

I was most anxious leisurely to examine this port, where the ships of war take refuge during the hurricane months, but, the Wellesley being at anchor close to the shore, on a precipitous bank, I deemed it proper not to risk the ship for a night in that situation.

I brought off samples of the strata, and various petrifactions which seem to have been the stems and branches of trees, replaced in appearance by phosphate of lime. There are samples, too, of animal petrifaction, enclosed in stone, which I should otherwise have considered of igneous origin; and charcoal is imbedded in masses of slate. Some of the limestone which I saw in the yard is of the purest white.

Although chiefly of volcanic materials, all these

islands are worthy of being examined by really scientific persons appointed by Government; and I have no doubt the cost would be amply repaid. The surf on the beach prevents landing from boats; therefore, any inspection of a minute nature must be prosecuted on shore. We weighed at sunset, and stood towards Montserrat. An eclipse of the moon shortly after commenced, and the sky being very clear, we had a beautiful view of its progress, and observed with interest the great expanse of the arch of the earth's shadow compared to the disc of the moon. The breeze was very gentle, and having but a short distance to go, we proceeded under topsails alone.

MONTSERRAT.

March 9th. — At sunrise we were about eight miles from Montserrat, and by nine o'clock had approached to within two miles. Volcanic ashes, stones, and red and white earths, are heaped alternately in masses, pervaded by tortuous strata of lava. There appear to be bands of ironstone, if a judgment can be formed from its colour; but the surf ran so high that it would have risked a boat to attempt to land to get specimens.

The most lofty mountains have their summits truncated, resulting probably from exhausted craters. The anchorage is reported to be deep and bad, and as the surf seems to run on the beach to leeward, with as much violence as on the weather side, (the

Wellesley being in tow of the Scourge,) I did not consider there was justifiable cause to attempt to anchor.

The cultivation of sugar seems to have diminished, for there is comparatively little growing cane; and windmills which suffered by the hurricane last year have not been repaired.

I have since had an opportunity of again visiting this island, which enables me to add the following particulars :—

The land has almost entirely been sold for the Government Taxes, which still continue at the amount imposed in prosperous times. In the official account given by the President, and published in the Blue Book, it appears that an estate, taxed as of £10,000 value, was brought to the hammer for £100 arrears of taxes, and sold for £116, which excess of £16 did not suffice to pay the appraiser and auctioneer. Another estate valued at £8000 was put up for £80 of taxes, and sold for precisely that sum, the appraiser and auctioneer getting nothing for their trouble. A house and garden in Plymouth (the capital) rated at £450, was sold for £45. Where Government is meant as a protection for the rights and properties of subjects such proceedings are shameful. I am not aware that in Turkey or Algiers more summary or stringent measures could be resorted to.

Surely the truthful reports of the President of Montserrat, have never been read by those who can remedy the monstrous injustice of continuing to tax

estates which make no return, and houses whose
proprietors have no income, at the full amount that
was levied before the universal ruin which has been
occasioned—not by the emancipation of the Slaves—
not by their unwillingness to labour, but by the
fatal blow that has struck down all Colonial interests throughout the West Indies, by the iniquitous
admission of slave-grown sugar in competition with
that of our highly-taxed West India Colonies—
thus encouraging Slave-owners and Slavery, and
rendering the vast outlay of treasure, and waste of
life, in the suppression of the Slave Trade, unavailing. With one hand we give a bounty to our
seamen for the capture of slave vessels, whilst with
the other hand we hold out a premium for the
thraldom of Slaves. Would it not be more judicious, and more effectual in the suppression of
Slavery, to remit all taxes on food, and on the production of sugar, in our free Colonies, and even to
apply the cost of our African squadron, and the
salaries of the numerous Slave Commissioners, to
enable free sugar to come in competition with, perhaps to undersell, that produced by Slave labour.

I have had sufficient means to confirm my observations thus far, and having sounded the Bay of
Montserrat, and found a bank whereon there is as
good anchorage and shelter as under any island in
the West Indies, I resolved to inquire particularly
into the state of this island, to which the frank
and manly reports of the President had called

my particular attention; not that the case of Montserrat is solitary, but because I felt I should waste less time in arriving at facts.

The result of my information is that the same onerous taxes continue to be exacted; and those who are unable to pay them, are uniformly levied on by the forced and ruinous sale of their property. In this way one estate of 800 acres had just been sold for £29; and another of 200 acres for £3.

The whole population, judging from their dress and the appearance of their houses, seem to be in a state of great destitution. Indeed how can it be otherwise. Food, which in England is introduced free of impost, is here subject to heavy Custom-house dues. The most extraordinary feature attending such exactions is, that the consumption being controlled by the circumstances of the impoverished community, the whole amount of Government taxes levied (though a grievous burden) does not defray the expense of the Custom-house department, which draws bills on her Majesty's Treasury for the *deficit* in their salaries !

I shall not dwell on detail, but turn to a more agreeable subject — namely, a trip to the boiling springs and sulphureous exhalations, which we witnessed after a ride of some miles over a road excessively bad, but affording many picturesque and beautiful views in our progress. All we passed over was volcanic. After clambering up as far as necessary, we descended into what had formerly been the crater, and having walked some distance over soft and spongy materials, perhaps the deposition of substances held in solution by boiling water, we arrived at the spring, which seemed to be heated much above the boiling point. The water appears quite black at its issue - probably from the impregnation of materials dissolved under great pressure. Copious fumes of sulphur exhale close by the spring, and might easily be condensed. The stones around are thickly covered with an incrustation, quantities of which it is said had been collected and shipped by some Americans, but no one on the island seems to pay any attention to the subject. Fifty yards from the boiling and sulphureous springs, there is a smaller rill of cold water, apparently a concentrated solution of alum, which might be turned to useful account; but this is unheeded, and lost in the hot and sulphureous current a short distance down the ravine. This cavity or crater is surrounded on three sides by lofty rocks, in some parts precipitous with an imposing, if not an awe-inspiring effect. Having satisfied our curiosity, we returned, partly by the road we came, and being less intent on the chief object of our journey, we looked at minor matters on the way, and here we saw growing the tree-fern, which I had been given to understand was only found as petrifactions, deeply imbedded in alluvial rocks, with the sole exception of some still existing in the South Sea Islands. They had no ripe seed, and the trees were far too large to remove.

We returned to the President's, and gave him many thanks for the horses which he had procured for us. The animals were not shod, although the roads were covered with sharp hard stones mostly of large dimensions. To my surprise the horse I rode did not make one false step, and, I believe, the others carried the whole party with equal security. The President had prepared a lunch, and shewed every mark of civility and attention in his power.

We weighed and proceeded on towards the N.W. point of Montserrat, from which a long but not dangerous shoal projects, and stood towards the small cone-shaped island called Redondo, which is also volcanic. It looks from the S.W. like a spinning top turned upside down, the waves having worn its lower circumference all around. Several shoals are laid down to the southward of this island, and therefore we stood to the W.N.W.

The weather is delightful, and temperature 75°, the wind being still in the northern quarter.

NEVIS.

March 9th.—In approaching Nevis the sun was declining, and it set before we reached the anchorage in Charlestown Bay, into which the Wellesley was towed by the Scourge; the character of the island is the same as the last-mentioned island, but the great mountain is of a more gentle acclivity, and the cultivation around its sides is more extensive than on mountains of the same description which we have before noticed. I did not land at Nevis, as I could not spare time for inquiry into circumstances which indeed could only have produced a repetition of that which I had already heard and seen. The Scourge was therefore ordered not to anchor, so I embarked at once, and steamed through the narrows between Nevis and St. Christopher, a bad channel even in a bright moonlight night and gentle breeze ; our course was then shaped for Barbuda, an island which has obtained the worst character of all the Caribbean range.

When we approached within twenty miles the steam was slackened in order to await daybreak, there being no survey of the shoal water around this island. Fifteen miles run after daybreak, scarcely rendered its long low points visible, but the colour of the water and the multitudes of fish and seafowl seemed to indicate that they never were disturbed by man, and that caution was necessary in approaching the shallow water. We however steamed leisurely over soundings from 24 to 4½ fathoms, and at 10 A.M. anchored in the last-mentioned soundings ; the south part of the island bearing S. and the northern extremity N. by compass. A boat was sent on shore with Commander Wingrove and Mr. Cochrane, and landed through a heavy surf ; some negroes having assisted in hauling up the boat, it was placed in safety. A great lagoon is interposed between the landing place and the village.

Three white people reside on this island, one of them a clergyman, from whom some intelligence was received as to its state. It is held by lease from the Crown by Sir William Codrington, Bart. The island is wholly uncultivated by the lessee. About 600 negroes grow vegetables for their own consumption, but no fruit. These people, I conclude, enjoy a very idle life. Enquiry was made if they fished. They did so for their own use, but none was dried or salted, though abundance of salt might be procured from the lagoons.

They burn charcoal and lime for the supply of the neighbouring islands, and export a few cattle which run wild. There are deer, wild hogs, wild guinea fowls and other animals and birds. There are many turtle upon its sandy shores; but it seems there is no encouragement to catch them. I never saw an island which presented such an uninterrupted beach of sand for the resort of that amphibious animal.

Mr. Cochrane brought off some samples of the coralline rock of Barbuda. It seems even a more recent formation than Barbadoes; the shells being still more perfect and larger. It is evidently upheaved from the bottom of the sea, probably on an unbroken blister of volcanic matter. There is, however, no appearance of any igneous rocks, such as constitute the neighbouring islands.

Barbuda has neither harbour nor creek, but a

great shallow lake in its interior is connected with the sea through a small channel which permits the entrance of boats.

It is a remarkable circumstance that earth has not formed on the surface of this coralline formation by decomposition, although amongst the trees there is a slight covering of decayed vegetable substances. We weighed anchor, and directed our course back to Nevis, were we arrived at 10 P.M., having made a tour of the southern side of the island to avoid passing again through the channel.

Sunday, March 11th.—Day dawned upon the magnificent scenery above Charlestown. I longed to visit the hot springs, celebrated for their baths, to which there is attached a large hotel for invalids; but as these were not subjects of naval interest, the Wellesley weighed at 7 A.M., and it being calm, the Scourge took the ship in tow.

St. Christopher.

March 11th.—At 10 A.M. we arrived off the town of Basse-Terre, which is even more beautifully situated than Charlestown. The surrounding hills rise like an amphitheatre, and bear evidence of having been more carefully cultivated than the lands on any British island we had yet seen. The present crop, however, bears but a small proportion to that formerly grown. In the distance very lofty mountains add to the beauty of the scenery, one of which has its top shattered, evincing its active character in

distant ages. There are accounts by travellers of
the enormous crater it contains, said to be some
thousands of feet deep ; but this like other interesting
sights, it was necessary to pass by. We were towed
along shore within a mile of the beach, and I was
greatly struck by the appearance of the conical
mountains, obviously of volcanic origin, regularly
declining on all sides from a perfect apex, without
the slightest indication of the existence of craters.
I conclude they must have been forced up in that
form, for it is highly improbable that decomposition
could have produced their present shape.

St. Eustatius.

At noon we were close to St. Eustatius, a Dutch
island, the southern end of which consists of a lofty
extinct volcano. Stratified and tortuous beds of
various colours are now exposed in the precipitous
cliffs washed by the sea. At the S.W. corner the
rock is chiefly of a whitish colour, the strata running parallel to the declivity of the mountain. The
less elevated northern hills bear evidences of the
same origin. There seems to be little or no cultivation here—probably the acclivities on this side are
too precipitous.

Saba.

Another island, or rather mountain, is in sight
precisely of the same character as St. Eustatius.
I am grieved that the Scourge has made the signal

of having only twenty-four hours fuel, as it will oblige us to pass St. Martin's, Anguilla, and the British Virgin Islands, at a distance, in order to procure fuel at St. Thomas's. Saba is a volcanic mountain having a crater. The above mentioned islands, which we passed at a distance, were imperfectly seen, so that no observations can be ventured on their character.

St. Thomas.

At noon, on the 12th March, the Danish flag was saluted with twenty-one guns, which the fort returned.

The town has a very neat appearance; the houses and stores are large, newly and neatly painted, and denote, by their exterior, comfort and general prosperity. The Governor's aide-de-camp came on board to learn if I proposed to land, and three o'clock was appointed. I was met on the wharf by His Excellency, and a salute was fired from the same fort in the town. The Governor's residence is on a precipitous rising ground. His Excellency was extremely polite, and spoke English with much fluency.

Free trade, in the right sense, seems to flourish here, and its consequences to be highly beneficial to the population. There are one or two Danish vessels in port, but the United States' vessels carry on nine-tenths of the commerce. Besides the multitude at anchor which do not display their national ensign, there were eight under sail with their colours hoisted. This island seems to be the emporium for merchandize from all quarters.

Assuredly there is nothing to invite the settlement of any persons here but gain from trade. There is not a spot of cultivated ground on the island, nor is there a garden to be seen about the town. All their fruit and vegetables are brought from the island of Santa Cruz, forty miles distant.

The negroes are *free,* but they are required to employ themselves usefully, under the penalty of being forced to work, without remuneration, on the roads, and in the repair of forts or other public works. Incarceration on bread and water being the penalty of refusal or reluctance in the execution of these compulsory duties. The blacks are all well clad, and seemed healthy and cheerful, presenting a remarkable contrast to the free negroes in the British islands.

March 12th.—The Governor requested our company to dinner, which I declined, under the conviction that in this climate official dinners are the precursors of indisposition. The sun was so intensely hot that I was glad to return on board.

The town stands on three conical hills, which join at their bases, and wharves extend the whole length crowded with schooners and small vessels engaged in their respective kinds of commerce.

The shores that form this fine harbour, are constituted partly of volcanic materials, partly of a

kind of indurated clay-stone, which perhaps may
have been volcanic mud. In some parts it appears
like a conglomerate, split, cracked, and twisted in
all directions. There is nothing to indicate that
this island differs in its origin from those which evince
more clearly the direct effect of volcanic power.

March 13*th.*—At nine o'clock we weighed,
and made sail on our course to Jamaica, and there
being a fine breeze, the Scourge was taken in tow
to save fuel. We should have proceeded at an
earlier hour had we not been detained by a complaint from the seamen of a British vessel, in which
they had refused to go to sea, in consequence of the
imputed inebriety and outrageous conduct of their
commander. This vessel had been lying seven
weeks at St. Thomas', there being no authority to
settle the matters in dispute. The Captain of the
Wellesley took the trouble to investigate the affair,
but found the master too much intoxicated to give
any explanation. The crew complained of bad provisions and short weight. Both statements were
verified. The scales were false to the extent of
one-fourth. They also complained of ill-treatment
when the master was drunk, which happened
almost every evening. No redress could be afforded, as no instructions or authority are given to
the commanders of ships of war to interfere in such
cases. Injustice to the men, and injury to the
owners, were therefore left without remedy, as any
assumption of power without authority might be

brought forward by any minor legal practitioner in England, and represented as an arbitrary violation of the law. In a foreign port such scenes are disgraceful, not only to the individuals, but to the nation to which they belong.

The very unusual occurrence of a north wind has continued for ten days. Now the easterly trade has resumed its wonted course. The ship was steered S.W., to pass Beique or Crab Island, the nationality of which is disputed by England, and orders were once given to dispossess the Spaniards, who are said since to have placed a guard of soldiers there, although it is a worthless possession in every point of view. The eastern end is of a white material, like Basse Terre (Guadaloupe), Barbuda, Barbadoes, and other places already mentioned. Further to the westward the rocks on the shore are brown and variegated, like those at St. Thomas'. The course to the S.W. was continued during the night to avoid some real or reported dangers on the southern shore of Porto Rico. We were therefore so far off in the morning, that nothing could be clearly observed but the great mountain range in the interior, so high that their summits appeared above or were involved in dense clouds. These mountains are said to be unexplored to this day. If they are of the nature of the northern range of Trinidad, probably minerals might be discovered which would justify the appellation of Puerto Rico given by the discoverer. The energies of the Spanish inhabitants are so relaxed that subterranean riches will probably remain for ever undisturbed, unless Porto Rico falls under the dominion of a more energetic people.

The island of Mona, situated half-way between Porto Rico and St. Domingo, is uninhabited. It is low and covered with small shrubs ; the cliffs are of a sandy-white colour, probably of loose coralline formation, as the sea has made great havoc around it, much more than it could have effected on indurated rock.

From Mona we shaped our course in order to pass near St. Domingo, but night coming on we hauled off S.W. Had it not been that the squadron was waiting at Jamaica, I should have delayed to obtain a view of this island. The sea seemed alive with flying-fish, which spouted from the bows in all directions.

JAMAICA.

On the 17th March we saw the Blue Mountains of Jamaica, and on the following morning, it being calm, the steam power of the Scourge was again resorted to. We entered the harbour of Port Royal about nine o'clock, and took in moorings off the outer part of the Naval Yard, opposite to the Naval Hospital, which is a handsome building surrounded with verandas. The point which forms the anchorage is pleasing to the sight—the palm and cocoa-nut trees give it quite a tropical aspect.

H

March 18*th*.—The morning of the 18th was past in receiving the Commodore and Captains. Having dined alone, it being Sunday, we took a stroll in the evening on the promontory that forms the point of the long spit which gives shelter to Kingstown and Port Royal harbours. Although nine or ten miles in length it is entirely formed of sand, having numerous lagoons or swamps surrounded by mangrove bushes. The malaria was sensibly felt during our short walk. Two assistant surgeons in the Naval Hospital, which is situated within range of the malaria, both successively appointed by me within the last twelve months, were dead, and a vacancy occurred in the Imaum for a Lieutenant, who shared the same fate. This pestilential vapour is augmented by the exhalations from putrid matters in the town of Port Royal, which is the dirtiest place I ever beheld. The thermometer this day stands at 88°. The land breeze coming from the mountains at night, instead of adding to the salubrity of this place, cannot be enjoyed with impunity, as it produces fever by checking perspiration.

It is a matter of regret that the ships of war must be detained in this vile anchorage, by the necessity of holding courts-martial on the engineers and officers of the steam sloop Vixen. It, therefore, appears to me desirable next year to order the West India portion of the squadron to Halifax during the hurricane season, and there try delinquents at leisure in a cool and wholesome climate.

The Governor's aide-de-camp, Captain Grey, came on board to ask me to take up my residence at His Excellency's country-seat in the mountains, for which obliging invitation I returned my thanks, and offered to take the Governor round the island in the Scourge.

March 19th.—Desirous of paying my respects to His Excellency Sir Charles E. Grey, I set off before sunrise, accompanied by Captain Goldsmith and Lieutenant Cochrane, for Port Henderson in the galley, where we found the Governor's carriage waiting for us. The distance to the Government House is seven miles over an alluvial plain; the lower part of which is at times flooded by the sea. The inner portion gradually rises, but the road is most uninteresting, having high mounds of earth on each side, covered by large plants resembling those of pine-apples, growing apparently without moisture, in the greatest luxuriance. The thickets, which have never been cleared, are covered with plants of prickly cactus. The clear space along this road is called pasture land; but though I saw a number of lean sheep, I could not distinguish a blade of grass. The surface is arid detritus. It is said that rain has not fallen on this side of the island for three months. (See Appendix.)

His Excellency, the Governor, was extremely affable; and I was pleased to hear him converse on various subjects to which persons in authority rarely pay attention. His library is stored with an admirable collection of works on Jamaica, and on subjects connected with the interests of this and other islands in the West Indies.

I felt impressed that it was impossible so shrewd a man could fall into the errors of which the opposition press of Jamaica accuse him. Whatever may be the evils existing (and I believe they are numerous), a Colonial Governor, however inclined, assuredly has not the means of applying effectual remedies. He has the misfortune to govern men, soured by losses which, with some degree of justice, they ascribe to enactments of the Parent State; but which assuredly would not have produced the amount of existing distress, had not the careless manner of managing estates, and the extravagance of the planters themselves, involved them in pecuniary difficulties, from which the compensation for the emancipation of their slaves, was inadequate to relieve them.

The abolition of the differential duty on foreign grown sugar has proved a decisive blow to any prospect of profitably working the estates, as these do not now afford a sufficient return to enable the emancipated population to obtain an adequate reward for their labour. Consequent on this state of things is the loud complaint of the indolence of the negro population, whose poverty, if inferred from the rags they wear, is awful; however, that is no just criterion, for although they may not be able to purchase clothes, still "Pumpkin" is easily obtained for themselves and their families.

His Excellency declined taking the tour round Jamaica, on account of pressing business. I therefore resolved to proceed in the Scourge, for the special purpose of visiting Port Antonio, which I had learnt was a fine harbour, formerly frequented by ships of war, in the hope that I might find it a better rendezvous for the squadron than the unhealthy anchorage at Port Royal.

March 20th.—The Scourge's signal was made to cast off the moorings. I went on board accompanied by the Flag Lieutenant, and proceeded to sea before the gun was fired for holding the first court-martial on a Lieutenant of the Vixen, that on the engineers being reserved until the following day. The Scourge proceeded, and with the view of economizing fuel, consumed only three-quarters of a ton of coal per hour in eight fires; but no amount of fuel ought to be deemed extravagant in search of a more healthful anchorage.

About one o'clock we rounded the eastern end of Jamaica, which is low and covered with dense jungle, though the hills at a distance bear evidence of culture; and on rising grounds are houses sufficiently neat and large to warrant the conclusion that proprietors reside on those estates.

At sunset we entered Port Antonio, too late to see the village or surrounding objects.

March 21*st.*—On the morning, the dawn of day presented a most beautiful landscape of hills, over-topped by lofty mountains, and a neat village on the beach, having a spacious church sufficient to hold ten times the number of people who reside within sight. On the western side of this bay, connected by a narrow but deep entrance, there is a harbour quite land-locked, capable of accommodating nume-rous vessels of any size.

The promontory which forms this anchorage, is of moderate height, covered with jungle. The people of the village are said to be healthy, as well as the black troops; but the fact of black soldiers being quartered here, in my opinion, renders the salubrity of the situation doubtful. The night air came down from the mountains with a chilling cold, and at times there was heavy rain, which occurs frequently on the northern side of the island, although, as already mentioned, no rain had fallen on the country near Port Royal for several months.

On inquiring of a person who called himself a pilot (though there is no occasion for a pilot here) how the planters were getting on, and how the negroes performed their work on the estates ; he said there were abundance of able labourers, but that the wages offered were not sufficient to maintain them ; hence the neighbouring lands, once extensively cul-tivated, are abandoned — nothing like sugar cane now appears around this village, nor within sight of the anchorage. We could get no vegetables, though they are said to be abundant on the market day, nor any fruit, but cocoa-nuts and plantains, which grow without care or culture. On one bunch of the latter there were 109 full-grown plantains. No fish could be got, although plentiful along shore and in the harbour.

The pilot told me that he had a diploma from the Duke of Manchester, and that he ought in justice to be the Harbour Master, but that a non-resident held the office. I inquired what the dues were which his rival received for doing nothing, and he told me he would furnish me with a list, which he did on the following day. Such exactions on commerce do much harm, far beyond the amount extracted, even if it were placed to the account of revenue, instead of being pocketed by useless individuals.

Having examined the western harbour, which is capacious and sheltered, and procured some specimens of the surrounding rocks, which are calcareous and full of shells, we departed at 9 A.M. and steamed along shore to the N.W. The cultivation seemed almost abandoned for the first twenty miles. However, when we got past Antonio Bay, the hills began to exhibit the characteristics of Jamaica in former times. There is one large estate extending over several hills, entirely under culture, and covered with most luxuriant looking sugar-canes. It would be desirable to learn the circumstances of the management which renders it so conspicuous, and such a contrast to other estates in the island.

About noon this day, a water-spout appeared a short distance to windward of the Scourge, partly rising from the sea, and part descending from a dense cloud. It seemed to revolve with great rapidity, in a direction contrary to that of the sun in the northern hemisphere. The Scourge's course was altered to avoid being involved in its vortex, and a gun was fired to create a concussion, and to ascertain its effects, if any, upon the phenomenon. The velocity of the rotation certainly ceased soon after, and the upper or pendant portion disappeared, and in a few minutes the lower column descended into the sea.

We continued to steam along shore, passing Annotta Bay, Ochorios, Dry Harbour, and lastly Falmouth, where the night commenced. The steam was then stopped, in order to creep slowly along shore under sail.

March 21*st.*—All the bluffs, where rock was exposed during this day, were of calcareous formation. Some estates seemed admirably cultivated ; a greater number were much neglected, and a large proportion had the appearance of being wholly abandoned to the production of a very tall grass, which on this rainy side of the island looked so green that at a distance it might be mistaken for sugar cane.

March 22*nd.*-- Beyond Pedro Point, being to leeward of the island, the water became quite smooth, and numerous fishing-boats take advantage of this circumstance. We called one or two alongside, but we were not tempted by its appearance to purchase the fish they had caught. The bays are everywhere suited to landing. A boat went on shore to bring off samples of the rock, which is limestone. The bay between North and South Capes Negro has a beach of white coralline sand, and along shore, a thicket of mangroves. One of the fishermen told us that plenty of turtle resorted to this beach, but they took no heed of them, as they could not get them to market. From these people we now procured some fish, but they refused money, saying it was of no use to them, and asked for a bit of salt pork.

Looking up this bay, which seems from the colour of the water to be shoal, the mountains in the interior of the island divide and form the channel of a river, which no doubt occasions the coloured water in this bay. We ran along the shore at the rate of five miles per hour, using only 12 cwt. of Welsh coal. The coal we got of this description at Port Royal is of a curiously crystallized formation, and is the best 1 have ever seen. The sending out of such coal is creditable to the Admiralty; for, judging from facts within my knowledge, considerable resolution is required to overcome the influence of those who have had coal to sell, and to counteract the interest of those who receive it from contractors.

At 11 A.M. we rounded the S.W. Point and immediately encountered the sea breeze, of which we had been for some time deprived. Negroes are settled upon the beach in considerable numbers, and exist with little exertion on fish and yams.

At Cape Negril the boat was again sent for samples of the rock, where it appeared recently fractured. Though scarcely a mile from the shore, which is low, we sounded and found eighty fathoms. The boat returned with specimens of the calcareous rock, which contained madrepores, a large univalve, and several bivalve shells.

The southern portion, from Savanna to Portland, has numerous shoals and rocks, of which there is no chart to be depended on; and being tired looking at the same objects and formations during 150 miles, the Scourge was kept under steam all night, and we arrived at Port Royal in the morning. (*March 23rd.*) Here we found the court-martial on the engineers of the Vixen still going on.

March 24th.—I went this morning to visit the Naval Yard and Hospital. All in the Yard, except the drains, is in good order, but the cost of artificers and labourers might be greatly diminished. The Yard is nearly on a level with high water, and, consequently, the surface is flooded during heavy rains. I went to the top of a tower (which has been erected at a cost of £1300 to accommodate a clock worth £50), not only that I might judge of the thoughtless extravagance of this building, but that I might have a view of the swamps on the promontory from which pestilential vapours exhale.

The burial-ground is in the same quarter, directly to windward of the Yard and Hospital. Some of the officers, whose friends or acquaintances were interred there, visited their graves and found coffins fresh in appearance exposed to view, the sand having been swept off by the tropical breeze.

It is a painful sight to observe the total absence of sanitary superintendence of this resort of the Naval service, and of the troops who ought to constitute its garrison. These being removed to quarters in the mountains, the duty of providing such sanitary measures for the Naval department has been urgently represented to the Admiralty.

The Hospital is clean, capacious, and well constructed, and is in one of the best situations for ventilation, if pure air could approach it. I have already mentioned the malaria produced by the lagoons, which pervades the whole promontory called the "Palisades." To this malaria, affecting the Hospital, are added the emanations from the filthy enclosures attached to the huts of the negroes to windward. The quarters of the assistant surgeons have a negro wash-house, a piggery, and a court full of putrid matter, close under their windows. Every one that has occupied this den has been ruined in health, or died. I have, therefore, taken upon myself the responsibility to order a spare ward in the Hospital, to be appropriated to the assistant surgeons. The lower range of apartments is wholly unoccupied, they being far more subject to the injurious effects of malaria. The floors are within a couple of feet of the stagnant water under the wards, which are, in my opinion, quite unfit for the reception of patients under any circumstances. They might be useful if converted into store-rooms for the Hospital, and the detached buildings (now appropriated as stores) converted into dwellings for the assistant surgeons.

Water is brought from a distance of fifteen miles, whilst that from the extensively-slated roof of the Hospital runs to waste, there being no tank to receive it.

It is strange, that whilst the military barracks are now generally removed to healthy situations, an auxiliary hospital for convalescent naval officers and seamen, situated in the mountains, is ordered to be sold.

The evening is the best time to depart from Port Royal if ships can get out beyond the reefs before dark, that is, if they are bound to the eastward. Having a steam vessel to tow the Wellesley, we were enabled to avail ourselves of the land wind from the mountains. In the morning a calm intervenes between the setting in of the tropical breeze, and of this we took advantage by proceeding in tow towards Cape Nicola Mole, where we anticipated that we should be able to stand to the northward under sail. A light sea breeze however set to the eastward, under the lee of St. Domingo, therefore the steamer was continued ahead.

I feel inclined here to note down circumstances which have been called to my observation, by official reports and other documents, setting forth numerous complaints of the depreciated value of produce—the want of return for capital invested in the staple commodities of the island—the reluctance of the negroes to perform any duty—the absence of trade—the universal impoverishment of the higher classes, and the wretched destitution of the labouring population. There is much truth in all these complaints, but I hesitate to admit to the full extent the causes to which they are ascribed, or the validity of the remedial measures proposed.

No doubt the generous and noble act, by which, in the reign of his late Majesty, Slavery was abolished, produced a prejudicial change in the economy of the sugar plantations, notwithstanding the large amount awarded to the proprietors, as the sums so paid were for the most part immediately transferred to mortgagees, leaving the proprietors in possession of the soil, but without the means of paying the expense of its cultivation by free labour. This is an evil which time has not remedied, and, of course, in the estimation of those who are, in consequence, losers, furnishes the pretext for imputing to the black population a degree of reluctance to labour far exceeding the reality. Those who pay a reasonable price for work, and are punctual in their payments, do not fail to get as many labourers as they require. I assert this not from any vague

hearsay, but from various unquestionable and
authentic documents, amongst which are the examinations taken by Committees of the House of
Assembly appointed to inquire into the causes and
difficulties alleged to exist in the cultivation of
estates.

Whilst the poverty of the planters and the destitution of the labouring population is so universal,
it seems most extraordinary on inspecting the
Custom House Returns to find almost every article
of necessary consumption brought from abroad,
paying high duties on entry; whilst the concession
of small patches of land to the negroes, whom there
is no capital to employ, would, if accorded, produce
food, and in a great measure dispense with such
injurious importations. Is it reasonable to instruct
the negroes in their rights as men, and open their
minds to the humble ambition of acquiring spots of
land, and then throw every impediment possible in
the way of its gratification? I perceive by the
imposts and expenses on the transfer of small properties, that a barrier almost insurmountable is
raised to their acquisition by the coloured population. I have learnt that small lots of Crown lands
are scarcely ever disposed of, though three-fourths
of these lands are still in the hands of Government.

It is lamentable to see the negroes in rags, lying
about the streets of Kingston; to learn that the
goals are full; the penitentiaries incapable of containing more inmates ; whilst the Port is destitute

of shipping, the wharves abandoned, and the store-houses empty ; while much, if not all of this might be remedied.

It may be asked, how is this to be effected ? and I answer—by justice, resolution, patriotism, and disinterestedness. Never can this wretched state of affairs be remedied, so long as taxes on the necessaries of life are heaped on an impoverished population ! Never can the peasantry raise their heads with a contented aspect, whilst every animate and inanimate thing around them is taxed to the utmost ! Not only is there a tax on land, and on the shipment of its produce ; on houses, outhouses and gardens ; on horned cattle and horses, but on asses and pigs ! and the severest penalties are enacted for concealment or suppression in the returns. Officials are employed for the gathering of pittances which (as in the case of the Custom House at Montserrat) do not defray the expense of collection !

The harbour dues and exactions are such that no vessel, when it can be avoided, is brought into the Port of Kingston ; consequently, though Jamaica is admirably situated, even more favourably than St. Thomas ; the former Port is abandoned, whilst that of the latter is filled with the shipping of all nations.

I have said that education and contentment are incompatible with the state of degradation, in which the coloured people (the most numerous class) are kept by the obstacles presented to their attainment of the comforts which are looked forward to by civilized man.

It must not, however, be inferred, that I disapprove of the very laudable endeavours to enlighten the minds of the coloured population—on the contrary, I desire to see means adopted to give *a much better kind* of education, and to as great an extent as practicable. I am in favour of instituting a college for the higher branches of knowledge at Jamaica, and on a scale sufficient to receive from the various islands all who are inclined to study. This cannot be effected unless aid is given, and there is no difficulty in pointing out where the means could be obtained for defraying this expense.

Whether such a useful undertaking can be entered into whilst all is in so great a state of disorganization throughout the British Islands in the western hemisphere, is a subject upon which it is unnecessary here to express an opinion.

HAYTI.

March 27th.—I regret that circumstances prevented my calling at Hayti, to ascertain the state of society in that quarter. Were I to judge from the total obliteration of all traces of agriculture, and almost of huts along the shore, I should conclude that all labour, beyond what is absolutely necessary for subsistence, has been abandoned. I saw but one house in the capacious bay, called Cape Nicola Mole,

and that was surrounded by a high wall; I suppose
to fortify that solitary abode.

Coralline formations appear to exist here as at
Jamaica, but to a much greater thickness; they
exhibit horizontal steps, as if each in succession had
been exposed to the action of the waves, and had
been elevated, at different periods, above the surface
of the sea. At various distances a substance of a
reddish colour interposes, which looks like indurated
clay.

Continuing under sail we passed by the island of
Inagua, and the rocky parts of the Caicos; all of
which are of coralline formation. There are inhabitants on both of these reefs, for so they may be called.
Numerous sea-birds resort to such parts as are
uninhabited, and the water around abounds with innumerable fish, which might, by industry, be converted
into a mine of riches. Salt is made in vast quantities at the island called Grand Turk, one of the
group;—none of them are capable of cultivation to
any extent. It is a remarkable fact, that these
coralline mountains of submarine origin, (some of
which attain a considerable elevation above water)
have sides so perpendicular that no soundings can
be had with 100 fathoms of line, close to the ledges
which surround them.

New Brunswick.

New Brunswick and British Guiana, two colonies of considerable magnitude, remain to be noticed, respecting which, as I have not visited them, I can only give extracts from Official Reports, transmitted to me by their respective Governors, in compliance with my request to be favoured with information on their state and prospects.

The former is entitled, " Report on the Sea and River Fisheries of New Brunswick, by M. H. Perley, Esq., Her Majesty's Emigration Officer at St. John,"—" laid before the House of Assembly, and ordered to be printed," 2nd January, 1850.

This Report is worthy of consideration by those who have the power to remedy evils which it minutely details : a concise sample of which may be quoted from p. 26.

" The absolute state of serfdom of the fishermen
" of Point Miscou has been particularly described,
" because there are other bodies of fishermen in the
" northern part of the province, who are held in
" nearly the same state of poverty and bondage.
" The more favoured inhabitants of New Brunswick,
" who dwell at a distance from its remote northern
" shores, will no doubt be surprised to learn that
" there are any of their fellow-subjects in the same
" colony who are even in a worse position than
" southern slaves, and of whose moral, physical, and
" spiritual wants less care has been taken."

It is evident from the publication of this Report,

(which like many other records of facts might have been suppressed), that the Governor and House of Assembly do not take blame to themselves. To whom then is it due? Or by what unavoidable cause are our strong and hardy fishermen reduced to a state worse than southern slaves? Is it partly by reason of flour, the chief article of food, being charged to them at 51 shillings per barrel in New Brunswick, whilst in the adjoining State of the Union it is sold at 17 shillings, and every article of food and drink in like proportion?—or, partly because of a premium of 243,432 dollars being given by the Government of the United States to their fishermen for last year's catch, whilst they imposed a duty amounting to 163,130 dollars on British fish sold in the very same market!

British Guiana.

The details contained in the Report of the state of this once flourishing province, are recorded by a Commission of twelve of the most distinguished inhabitants, appointed by the Governor, who most obligingly transmitted to me a copy entitled, " State and Prospects of British Guiana," printed by William Clowes and Son, Stamford Street, London, 1851.

Passing over preliminary observations, it is stated at p. 4 :—" Your Commissioners have drawn a " parallel of 20 years,"—" in the year 1829 there " existed 230 sugar estates and 174 coffee estates

" in British Guiana, of which almost the whole
" were in full cultivation; whilst on the 31st of
" December last, the colony numbered only 180
" sugar and 16 coffee estates, even nominally carried
" on."—"Names of the highest and most influen-
" tial have followed one another in the Royal
" Gazette with ominous rapidity, and the estates of
" men, formerly holding the first positions in the
" colony, have been successively brought to the
" hammer, and their owners absolutely beggared."

Page 5.—" A long succession of formerly culti-
" vated estates present now a series of swamps, over-
" run with brushwood, and productive of malignant
" fevers."

Page 6.— " The question" (of prosperity or ruin),
" rests upon the line of policy to be pursued by the
" British Government with regard to the West
" Indies in general; for your Commission are of
" opinion that few estates (if any), however favour-
" ably situated can long continue to resist the
" tendency of the present free-trade measures."

Page 14.—" In the present state of trade, the
" tonnage dues and pilotage fees on all vessels
" coming into port are felt to be excessive, and the
" amount of additional fees paid to the harbour-
" master is a subject which urgently demands legis-
" lative supervision."

Page 16.—" A great trade in provisions is carried
" on with the United States of North America,
" upon which this colony is almost dependent for
" food." All of which is taxed !

Mosquito Territory.

One ill-omened possession—if possession it can be called—remains to be noticed, namely, the pestilential swamp or desert, called the Kingdom of Mosquito, to the deadly atmosphere of which, most of the sickness experienced in the Squadron during my command is to be attributed, and the consequent impunity with which the slave-trade to Cuba was carried on.

No sooner had a ship of war arrived at St. Jean de Nicaragua, and communicated with the shore, than fever commenced; and in one instance the crew of the vessel sent to relieve the station had to weigh the anchor of the other, as there were not a sufficient number of unattacked or convalescent persons on board to get her under weigh in order to her return to Jamaica.

Nicaragua is one of the rival routes through which it is proposed to carry a canal from the Western to the Pacific Ocean. Hundreds of thousands of labourers would inevitably perish in executing such an undertaking; which, if ever completed, it is to be hoped will not produce on the adventurers by such a course the deadly effects of the malaria of that region, too often fatal even without the predisposing cause of bodily exertion.

APPENDIX.

It would swell this Appendix to a volume were I to insert even brief quotations from all the numerous documents in my possession in proof of the truth of the facts set forth in the preceding Notes. Therefore, with few exceptions, it will only be hereafter and in the event of any of my statements being contradicted, that I shall publish corroborative documents.

NEWFOUNDLAND.

In the case, however, of this most important island, in addition to the Statistical Chart published herewith, I beg to refer to the document entitled "Report of the Select Committee appointed to inquire into the state of the Fisheries on the banks and shores of Newfoundland." (Ordered to be printed by the General Assembly, in April, 1845). This Report contains a most interesting detail which I cannot too strongly recommend to the serious consideration of the public. It is impossible not to share in the surprise which the Committee express in the following paragraph: "Your Committee cannot avoid expressing "their great surprise at the utter indifference and neglect "with which the British Government now appear to regard "these fisheries, more particularly when they observe the "assiduity and perseverance with which the French and "American Governments encourage them. The fact "cannot be hidden,—it must be acknowledged that the "great deep-sea fishery of Newfoundland has ceased to be "a nursery for British seamen, and has now become the "most valuable nursery for both the French and the "Americans."

The same Report refers to a Committee of the House of

Commons on the depression of the Newfoundland trade,
so far back as 1817, and quotes the following passage from
the Report of that Committee :—

" It appears also to your Committee, that the trade
" itself has experienced a serious and alarming depression.
" The causes from which this has arisen will require, in
" the opinion of your Committee, in the ensuing session of
" Parliament, a much more detailed and accurate investi-
" gation; but enough has been shewn, by the testimony
" of respectable witnesses, to prove, before this House
" separates, that the fisheries will be most materially
" injured, the capitals embarked in them by degrees with-
" drawn, and the nursery for seamen, hitherto so justly
" valued, almost entirely lost."

On the above passage, the aforesaid Colonial Committee
make (among other remarks) the following :—

" Notwithstanding this strong representation on the part
" of a Committee of the British House of Commons, the
" subject has not since been taken up by the Government.
" No relief or support has been afforded from that
" period to the present." * * " The British fishery is
" now confined to an in-shore fishery, prosecuted in punts
" and small craft, leaving the deep-sea fishery on the
" great bank, and the other valuable banks and fishing-
" grounds, altogether in the hands of the French and
" Americans."

Thus we see that the investigation by the British House
of Commons in 1817, and which was expected to have
been resumed in the following session, was not then resumed; and never was resumed in the 28 sessions which
ensued between 1817 and 1845, when the Colonial Committee made their above Report : nor has it been renewed
in any of the five sessions which have since elapsed. In
the case of this once most important nursery for seamen, as
in that of other subjects connected with the interests of the
Navy, investigation has been checked, and consequently there

is no abatement of old evils, nor hindrance to the introduction of new ones. And here, having alluded to the Naval Service, I beg leave to repeat the opinion I expressed in a former pamphlet, published in 1847 [Observations on Naval Affairs, p. 29] : — " Nothing can disengage the " Service from the drags by which its improvement is " retarded, but the revival of the Commission of Naval " Inquiry, quashed by a vote of the House of Commons " on the 1st of March, 1805."

I repeat that the Newfoundland Fisheries ought to be considered by the British Government, as they are by the Governments of France and America, namely, as nurseries for seamen in time of peace to be available in case of war; for, as truly set forth by Adam Smith, in his " Wealth of Nations," b. iv. c. 2. "The defence of Great Britain " depends very much on the number of its sailors and " shipping. The Act of Navigation therefore very properly " endeavours to give the sailors and shipping of Great " Britain the monopoly of trade in their own country, in " some cases by absolute prohibitions, and in others by heavy " burdens on the shipping of foreign countries.—As defence " is of much more importance than opulence, the Act of " Navigation is perhaps the wisest of all commercial " regulations in England."

If so, why has it been repealed?

ISLAND OF CAPE BRETON.

In my " Notes" (p. 25) on visiting Sydney, the capital of the island, I briefly noticed the beauty of the locality and the capability of the surrounding country for the production of all that man requires for his subsistence or is attainable from our native land. Reflecting how injuriously the interests of this kingdom are affected by the tide of emigration flowing more largely in other directions than to our own as yet under-peopled colonies, and especially how beneficially a portion of the tens of thousands who annually expatriate themselves, might devote their industry to the cultivation of the valuable possessions alluded to, I think it relevant to insert an extract from my reply to an address presented to me by the inhabitants of Sydney, in which my estimation, formed by actual observation, of the advantages, present and prospective, of these possessions, is more particularly expressed.

" The importance of the geographical position of this island has long been known to me, but I confess I did not give credit to its manifest agricultural capabilities; to the amenity and beauty of its scenery; neither did I reflect on the unparalleled advantages of its internal communication, extending over a littoral of an hundred miles, capable of being converted to means of transit by a short canal at a cost less than is usually expended on 500 yards of an English railway.

" That which at present would most speedily advance the prosperity of this island seems to me to be encouragement to the hardy fisherman and the industrious agriculturist by a fair market, whereby they could obtain a merited reward for their labours. The fisheries I understand are under the consideration of the British Parliament; the agricultural interests can be most readily benefited by an influx of population (which the parent state can advantageously afford), and that undoubtedly would not be delayed were the nature of these most important possessions of the Crown known to the public.

" All that Scotland, my native land, can produce is successfully cultivated, and is now actually in growth around this spacious harbour, at the entrance of which the grounds of the Mining Association would do credit to any country.

" I was gratified to see the manifest advantages derived in Newfoundland by the judicious establishment of colonial premiums for the production of the higher orders of grain, —for the rearing of cattle, and for the orderly arrangement and management of farmsteads. In no quarter that I have visited have I seen more care in the production of manure by the judicious combination of perishable substances with peat—beneficial arrangements which have been assiduously promoted by his Excellency the Governor of that island.

"I hope I shall be able to learn that similar measures are resorted to here, on which I shall have great pleasure in congratulating you on my return. At present, I regret to say, that I have seen an obvious laxity of industrious habits, arising, I believe, from want of stimulus to exertion.

"*Wellesley, Sydney, August 20th,* 1849."

Nova Scotia.

I need not enter into statistical observations which can be found in the annual Blue Book and other published documents, and from which every intelligent reader can draw significant inferences; but have merely to express my regret that there are no accessible means with which I am acquainted, of bringing petitions for the redress of grievances agreed upon and forwarded by the colonists, with the counter representations of resident officials, *jointly* before the public, so as to come at once to a decided judgment as to how far, if to any extent at all, the grievances alleged by the former, are disproved by the latter.

Prince Edward's Island, which was formerly attached to the government of Nova Scotia, has been advanced to the rank of a distinct colony. The soil is the best adapted to agriculture of all the northern colonies: yet is still comparatively a wilderness producing little grain. Though situated in the middle of the North American Fisheries, no advantage is taken of its favourable position. One circumstance however, is worthy of remark, that all Crown land having been granted in large lots in the middle of the last century, so that there is not an acre of Crown land remaining, there is still on the Governmental establishment a paid Surveyor of Crown lands, who and whose predecessors, have enjoyed the sinecure for an hundred years!

BERMUDA.

Having mentioned with pleasure the judicious conduct pursued by the local authorities of Bermuda in all affairs under their direction, it is with great regret that I feel myself bound to notice injurious consequences arising from measures beyond their control. The principles of free-trade are admirable when they can be reciprocally carried out, but unfortunately, such is not generally the case, and assuredly not in the instance which I am about to relate. The trading vessels of Bermuda are manned (with the exception of the master) by the coloured inhabitants, and these are seized, although under the British flag, at the moment a Bermudean vessel enters a slave-port of the United States, and are imprisoned during the whole period the vessel remains in port. Meanwhile the master is compelled (at the expense of the owners), to hire people to execute the duty which his crew were bound to perform. The owners too are charged with the cost of the maintenance of the imprisoned crew for whatever period they are confined; and it is publicly notified, that unless such cost is paid, the crew will be *sold* to defray their expenses! I possess a copy of an official caution, (affixed in the Court-house at Bermuda), warning the masters and crews of Bermudean vessels of the treatment to which they would expose themselves by entering the slave-ports of the United States! Ship-building, until lately the staple manufacture of the island, has suffered to such an extent, that cedar-trees (as I was informed by a most respectable land-proprietor), had fallen 75 per cent.

TRINIDAD.

To shew the ruin to which this most important island is
reduced—the most important of all the British Islands in
the West Indies—possessing unrivalled fertility of soil and
to the greatest extent—I insert an extract from the advertisements in the Port of Spain Gazette of December 31st,
1850, of properties for sale, under order of the "Court of
Intendant" for "Ward Rate," shewing the fearful extent to
which immense properties are sacrificed through the inability of the proprietors to pay even the smallest Government dues. Nor is it more than an average specimen of
the like terrible processes going on for a lengthened period
before, and undoubtedly still going on there and throughout the islands.

"The Port of Spain Gazette, Trinidad, Tuesday, Decem-
"ber 31, 1850.

"*Court of Intendant.*

"Public Notice is hereby given, That on Saturday the
"1st day of March, 1851, at the hour of Eleven o'clock in
"the forenoon, there will be put up and exposed for sale
"by Public Auction, by order of the Tribunal of Intendant,
"at the doors of Government House, the following under-
"mentioned Properties, situate in the Ward of Santa
"Cruz. These Properties will be put up for sale at the
"upset price of the several amounts due for Ward Rate,
"according to the subjoined list, and in default of pur-
"chasers will be adjudged to the Queen, and dealt with
"as any other Crown Lands in the Colony.

"THOMAS F. JOHNSTON,
"*Escribano.*"

Here follows the particulars of the estates so ordered to
be sold (herein only enumerated.) And in the same news-

paper are 14 more similar advertisements of sales in different districts. The summary of the whole is :—

Districts.	Number of Estates.	Districts.	Number of Estates.
Santa Cruz	24	Couva	12
Carapichaima	4	Erin	21
Mucurapo	16	Maraccas	23
Cedros	4	La Brea	31
Chagachacare	2	South Naparima	3
Chaguaramas	30	Couva (2nd list)	30
Point-a-Pierre	39	Diego Martin	17
Las Cuevas	8	ESTATES	264

On which estates there are 199 houses.

JAMAICA.

I shall quote a few items from the examinations taken before a Committee appointed by the Honourable House of Assembly to inquire into the present state of the agricultural interest—printed in 1846 by Jordon and Osborn, Spanish Town, Jamaica—wherein it appears, by the evidence of the most intelligent witnesses, that there is in reality no want of labourers, if constant employment and regular payment could be afforded to encourage exertion; and further, that much better crops would be produced, if funds existed to render the mountain streams available by conveying them to the parched districts—a measure which the planters neglected to resort to in the time of prosperity, and which they cannot now effect by reason of their poverty.

George Price, Esq. being sworn, says :—" If capital were invested in conveying water into the many dry districts, it would secure the plants, and a large first and second ratoon." " I have examined this matter carefully, and I repeat that the conveyance of water to these districts and the carrying out of this system would ensure immense crops where they are now small and uncertain." "It is impossible that in a mountainous country like Jamaica, abounding in springs, there can be a want of a sufficient supply of water." "I repeat that it is a simple question of capital, and that water could be conveyed at such an expense as would still give a large and sure return to the proprietors." "There is nothing that would so much tend to improve the industrious habits of the people as a well-devised scheme of irrigation."

Other gentlemen gave evidence to a similar effect. I do not refer to the opinions expressed in the above document as a remedy for the evils mentioned in the "Notes," but as pointing to a measure calculated to remove a portion of the existing distress.

It is, however, in vain to hope that any enterprise of magnitude will be undertaken by the planters of Jamaica in their present impoverished and dispirited condition. Even the protection of their own health is not a sufficiently powerful incentive to induce them to attempt the removal of the causes of disease with which they are surrounded. On the part of the Navy I used all the means in my power to guard against these pestilential influences. So offensive were the effluvia of Port Royal, that I not only felt it my duty to address the Admiralty on the subject, and to complain of the abomination in a Letter to the Mayor of Kingston, but I removed the flag-ship (though nearly a mile from the town) to the outer anchorage, in order to escape the malignant exhalations perpetually arising from that sink of corruption. I subjoin a copy of my letter to the Mayor of Kingston, (written a few months previously to the breaking out of the cholera), calling his attention to the unparalleled filthiness of the place, which unquestionably was the cause of the fatal distemper which ensued, and has swept from the scene tens of thousands of the inhabitants, and is still raging.

"Wellesley, March 21, 1850.

"SIR,

"I communicated to the Admiralty, on my visit to "Port Royal last year, the great injury arising to the "seamen consigned to the hospital, not only from the "numerous small enclosures which intervene between that "establishment and the trade wind, but from the exces-"sively dirty condition in which these and the whole of "the streets are suffered to remain, apparently without the "slightest attention being paid to the sanitary condition "of the town, or to the prejudice that may be incurred by "the seamen of her Majesty's ships who of necessity fre-"quent this place.

"It is my anxious wish that a sanitary commission "should be named, to which I have desired Commodore "Bennett, with your sanction, to appoint such officers as "may be in port at that time, to aid in an investigation, "with a view to remedy these evils. Never have I seen, "in the whole course of my life, a place so disgustingly "filthy, or which could give so bad an opinion to foreigners "of British Colonial Administration, as the town of Port "Royal.

"I have, &c.

"DUNDONALD,

"Vice-Admiral.

"The Honourable Hector Mitchell,
"Mayor of Kingston, Jamaica."

Although I have expressed an opinion that our Colonies

generally, under a suitable system of Government, would increase the power and prosperity of Great Britain, I am sensible that there are physical drawbacks to the amount of advantage which might otherwise be derived. The limited area of Britain seems to afford too small a basis for its vast domestic and colonial superstructure. Colonies, so numerous, so distant, and in such widely different directions, contrast most unfavourably with the acquisitions of the United States, consisting of adjoining provinces, which they have taken into the Union, and by which they have more than doubled their extensive territory. The long arm of the Union now embraces the neck of Mexico by the connecting canal with California, which new possession is rapidly advancing to the importance of a long-settled country, possessing all the advantages of a state of maturity without any of the decrepitude of age. A vast custom-house is in progress of erection : docks have already been commenced for the repairs of shipping, more capacious than any possessed by the maritime States of Europe— prospectively anticipating an extensive if not an exclusive commerce with India, China, and the Islands of the Pacific Ocean. All is progressing throughout the whole extent of this vast Republic (united by canals, railroads and electric telegraphs), to a height of prosperity and greatness, to which the state and prospects of the Possessions, forming the subjects of the preceding " Notes," present a contrast painful to contemplate.

NORMAN AND SKEEN, PRINTERS, MAIDEN LANE, COVENT GARDEN.

9 781108 054065